# Natural Bioactive Compounds from Fruits and Vegetables as Health Promoters
## *Part II*

### Edited by

**Luís Rodrigues da Silva**

*CICS – UBI – Health Sciences Research Centre*
*University of Beira Interior*
*Covilhã*
*Portugal*

&

**Branca Maria Silva**

*CICS – UBI – Health Sciences Research Centre*
*University of Beira Interior*
*Covilhã*
*Portugal*

# Natural Bioactive Compounds from Fruits and Vegetables as Health Promoters

Authors: Luís R. Silva and Branca Silva

ISBN (eBook): 978-1-68108-243-1

ISBN (Print): 978-1-68108-244-8 © 2016, Bentham eBooks imprint.

Published by Bentham Science Publishers – Sharjah, UAE.

First published in 2016.

advertisements or ideas contained in the Work.

## *Limitation of Liability:*

In no event will Bentham Science Publishers, its staff, editors and/or authors, be liable for any damages, including, without limitation, special, incidental and/or consequential damages and/or damages for lost data and/or profits arising out of (whether directly or indirectly) the use or inability to use the Work. The entire liability of Bentham Science Publishers shall be limited to the amount actually paid by you for the Work.

## General:

1. Any dispute or claim arising out of or in connection with this License Agreement or the Work (including non-contractual disputes or claims) will be governed by and construed in accordance with the laws of the U.A.E. as applied in the Emirate of Dubai. Each party agrees that the courts of the Emirate of Dubai shall have exclusive jurisdiction to settle any dispute or claim arising out of or in connection with this License Agreement or the Work (including non-contractual disputes or claims).
2. Your rights under this License Agreement will automatically terminate without notice and without the need for a court order if at any point you breach any terms of this License Agreement. In no event will any delay or failure by Bentham Science Publishers in enforcing your compliance with this License Agreement constitute a waiver of any of its rights.
3. You acknowledge that you have read this License Agreement, and agree to be bound by its terms and conditions. To the extent that any other terms and conditions presented on any website of Bentham Science Publishers conflict with, or are inconsistent with, the terms and conditions set out in this License Agreement, you acknowledge that the terms and conditions set out in this License Agreement shall prevail.

**Bentham Science Publishers Ltd.**
Executive Suite Y - 2
PO Box 7917, Saif Zone
Sharjah, U.A.E.
Email: subscriptions@benthamscience.org

**BENTHAM SCIENCE**

# CONTENTS

# FOREWORD

For centuries, humans have considered food only as an "energy" source for survival. Clarification of nutritional relevant components, as protein, fat, carbohydrates, minerals and vitamins, was determinant to understand metabolic needs, and to adjust consumption patterns. However, this oversimplified definition of food resulted in processed foods composed by mixtures of ingredients rich in these components, while diet is increasingly claimed as being responsible for the most common diseases of modern society: cardiovascular diseases, obesity, and cancer.

When we look upon food from this simplified perspective, it is as if we are regarding food without its "soul". Indeed, although being difficult to demonstrate causality between food and health, there is now appreciable epidemiologic evidence for the protective role of diets rich in fruits and vegetables, being the Mediterranean diet an interesting example. These foods have thousands of components without nutritional essentiality that have been neglected. The interest on these components has increased tremendously in the last two decades, seeking to identify the dietary bioactive components (*i.e.*, those that have a measurable impact on human health), their amounts, and availability. Simultaneously, it is also becoming clear that each one of these components has different effects and potencies when ingested alone or when taking its part in the complex network of molecules present in whole foods. These are amazing days for food scientists because we are closer to understand these bioactive compounds, while the consumer is following closely scientific advances, being increasingly interested in the health properties of foods.

The editors took an enormous and successful effort to assemble a huge variety of knowledge on different natural bioactive components in foods, bringing together experts working of different fields of food composition and health. Following a first volume on fruits, this second volume was written to provide readers with a comprehensive review of bioactive constituents in several legumes, nuts, seeds and cereals, from the most traditional ones, as rice or tomatoes, to emerging potentials in modern nutrition, as quinoa or coffee residues. This assembled knowledge allows the reader to get acquainted with the most promising bioactive compounds in different foods, understand the care needed to preserve their bioactivity during storage or processing, while revealing also the hidden bioactive potential of commonly rejected parts, as shells or seeds.

Therefore, this book is designed for food scientists, nutritionists, pharmaceuticals, physicians, food industrials, as well as for health-conscious consumers.

**José Alberto Pereira**
Mountain Research Centre (CIMO)
School of Agriculture
Polytechnic Institute of Bragança
Portugal
. &
**Susana Casal**
REQUIMTE / Bromatology Service
Faculty of Pharmacy
University of Porto
Portugal

# PREFACE

Plants have been widely used as food and medicines, since they provide, not only essential nutrients required for human life, but also other bioactive compounds which play important roles in health promotion and disease prevention, commonly known as phytochemicals. Moreover, in the recent years, the impact of lifestyle and dietary choices for human health has increased the interest in fruits and vegetables, as well as in foods enriched with bioactive compounds and nutraceuticals. In fact, epidemiological studies have consistently shown that the Mediterranean diet, characterized by the daily consumption of fruits and vegetables, is strongly associated with reduced risk of developing a wide range of chronic diseases, such as cancer, diabetes, neurodegenerative and cardiovascular diseases.

Phytochemicals are secondary metabolites present in fruits and vegetables in low concentrations that have been hypothesized to reduce the risk of several pathological conditions. There are thousands of dietary phytochemicals, namely flavonoids, phenolic acids, glucosinolates, terpenes, alkaloids, between many other classes of compounds, which present different bioactivities, such as antioxidant, antimutagenic, anticarcinogenic, antimicrobial, anti-inflammatory, hypocholesterolemic, hypoglicemic and other clinically relevant activities. The evidence suggests that the health benefits of fruits and vegetables consumption are attributed to the additive and synergistic interactions between these phytocomponents. Therefore, nutrients and bioactive compounds present in fruits and vegetables should be preferred instead of unnatural and expensive dietary supplements.

In this ebook, we provide an overview about the different classes of phytochemicals commonly found in fruits and vegetables, highlighting their chemical structures, occurrence in fruits and vegetables, biological importance and mechanisms of action. Volume 2 is dedicated to the study of several legumes, nuts, seeds and cereals.

**Luís Rodrigues da Silva**
CICS – UBI – Health Sciences Research Centre
University of Beira Interior
Portugal

# List of Contributors

**Adriana M. S. Sousa**    CICS – UBI – Health Sciences Research Centre, University of Beira Interior, 6201-506 Covilhã, Portugal

**Ana R. Nunes**    CICS – UBI – Health Sciences Research Centre, University of Beira Interior, 6201-506 Covilhã, Portugal

**Álvaro Peix**    Instituto de Recursos Naturales y Agrobiología, IRNASA-CSIC, Salamanca, Spain;
Unidad Asociada Universidad de Salamanca- CSIC 'Interacción Planta-Microorganismo', Salamanca, Spain

**Branca M. Silva**    CICS – UBI – Health Sciences Research Centre, University of Beira Interior, 6201-506 Covilhã, Portugal

**Carlos Albuquerque**    IPV - ESSV – Polytechnic Institute of Viseu, Higher Health School of Viseu, 3500-843, Viseu, Portugal

**Charu Lata Mahanta**    Department of Food Engineering and Technology, Tezpur University, Assam, India

**Cristina García-Viguera**    Research Group on Quality, Safety and Bioactivity of Plant Foods, Department of Food Science and Technology, CEBAS (CSIC), Campus University Espinardo, Murcia, Spain;
Phytochemistry Lab, Food Sci. & Technology Dept., CEBAS-CSIC, Murcia, Spain

**Diego A. Moreno**    Research Group on Quality, Safety and Bioactivity of Plant Foods, Department of Food Science and Technology, CEBAS (CSIC), Campus University Espinardo, Murcia, Spain;
Phytochemistry Lab, Food Sci. & Technology Dept., CEBAS-CSIC, Murcia, Spain

**Elsa Ramalhosa**    Mountain Research Centre (CIMO), School of Agriculture, Polytechnic Institute of Bragança, Campus de Sta Apolónia, Apartado 1172, 5301-855 Bragança, Portugal

**Encarna Velázquez**    Departamento de Microbiología y Genética. Facultad de Farmacia, Universidad de Salamanca, Salamanca, Spain;
Unidad Asociada Universidad de Salamanca- CSIC 'Interacción Planta-Microorganismo', Salamanca, Spain

**Gonçalo D. Tomás**    1CICS – UBI – Health Sciences Research Centre, University of Beira Interior , 6201-506 Covilhã, Portugal

**Isabel C.F.R. Ferreira**    Mountain Research Centre (CIMO), ESA, Polytechnic Institute of Bragança, Campus de Santa Apolónia, Ap. 1172, 5301-855 Bragança, Portugal

| | |
|---|---|
| **Joana S. Amaral** | REQUIMTE-LAQV, Departamento de Ciências Químicas, Faculdade de Farmácia, Universidade do Porto, Portugal; ESTiG, Instituto Politécnico de Bragança, Bragança, Portugal |
| **João A. Lopes** | iMed. ULisboa, Departamento de Farmácia Galénica e Tecnologia Farmacêutica, Faculdade de Farmácia, Universidade de Lisboa, Lisbon, Portugal |
| **Jorge M.G. Sarraguça** | LAQV/REQUIMTE, Departamento de Ciências Químicas, Faculdade de Farmácia, Universidade do Porto, Porto, Portugal |
| **José A. Pereira** | Mountain Research Centre (CIMO), School of Agriculture, Polytechnic Institute of Bragança, Campus de Sta Apolónia, Apartado 1172, 5301-855 Bragança, Portugal |
| **José Pinela** | Mountain Research Centre (CIMO), ESA, Polytechnic Institute of Bragança, Campus de Santa Apolónia, Ap. 1172, 5301-855 Bragança, Portugal; REQUIMTE/LAQV, Faculty of Pharmacy, University of Porto, Rua Jorge Viterbo Ferreira, n° 228, 4050-313 Porto, Portugal |
| **Lorena Carro** | Departamento de Microbiología y Genética. Facultad de Farmacia, Universidad de Salamanca, Salamanca , Spain |
| **Luís R. Silva** | CICS – UBI – Health Sciences Research Centre, University of Beira Interior, 6201-506 Covilhã, Portugal; IPCB – ESALD – Polytechnic Institute of Castelo Branco, Higher Health School Dr. Lopes Dias, 6000-767, Castelo Branco, Portugal; LEPABE – Department of Chemical Engineering, Faculty of Engineering, University of Porto, 4200-465 Porto, Portugal |
| **Mafalda C. Sarraguça** | LAQV/REQUIMTE, Departamento de Ciências Químicas, Faculdade de Farmácia, Universidade do Porto, Porto, Portugal |
| **Marco G. Alves** | CICS – UBI – Health Sciences Research Centre, University of Beira Interior, 6201-506 Covilhã, Portugal |
| **M. Beatriz P.P. Oliveira** | REQUIMTE/LAQV, Faculty of Pharmacy, University of Porto, Rua Jorge Viterbo Ferreira, n° 228, 4050-313 Porto, Portugal; REQUIMTE-LAQV, Departamento de Ciências Químicas, Faculdade de Farmácia,, Universidade do Porto, Portugal |
| **María Elena Cartea** | Group of Genetics, Breeding and Biochemistry of Brassicas, MBG-CSIC, Pontevedra, Spain |
| **Marta Francisco** | Group of Genetics, Breeding and Biochemistry of Brassicas, MBG-CSIC, Pontevedra, Spain |
| **Miguel Lopo** | LAQV/REQUIMTE, Departamento de Ciências Químicas, Faculdade de Farmácia, Universidade do Porto, Porto, Portugal |
| **Nieves Baenas** | Phytochemistry Lab, Food Sci. & Technology Dept, CEBAS-CSIC, Murcia, Spain |

**Pablo Velasco**  Group of Genetics, Breeding and Biochemistry of Brassicas, MBG-CSIC, Pontevedra, Spain

**Pedro F. Oliveira**  CICS – UBI – Health Sciences Research Centre, University of Beira Interior, 6201-506 Covilhã, Portugal;
Department of Microscopy, Laboratory of Cell Biology, Institute of Biomedical Sciences Abel Salazar (ICBAS) and Unit for Multidisciplinary Research in Biomedicine (UMIB), University of Porto, 4050-313 Porto, Portugal

**Ricardo N.M.J. Páscoa**  LAQV/REQUIMTE, Departamento de Ciências Químicas, Faculdade de Farmácia, Universidade do Porto, Porto, Portugal

**Sangeeta Saikia**  Department of Food Engineering and Technology, Tezpur University, Assam, India

**Susana Casal**  LAQV-REQUIMTE, Chemistry Department, Faculty of Pharmacy, Oporto University, Porto, Portugal

**Teresa Delgado**  Mountain Research Centre (CIMO) - School of Agriculture, Polytechnic Institute of Bragança, Bragança, Portugal
LAQV-REQUIMTE, Chemistry Department, Faculty of Pharmacy, Oporto University, Porto, Portugal

# Natural Bioactive Compounds from Fruits and Vegetables as Health Promoters
## *Part II*

# CHAPTER 1

# Bioactive Compounds of Legumes as Health Promoters

## Luís R. Silva[1,2,3,*], Álvaro Peix[4,5], Carlos Albuquerque[6], Encarna Velàzquez[5,7]

[1] *CICS UBI Health Sciences Research Centre, University of Beira Interior, 6201-506 Covilhã, Portugal*

[2] *IPCB ESALD Polytechnic Institute of Castelo Branco, Higher Health School Dr. Lopes Dias, 6000-767, Castelo Branco, Portugal*

[3] *LEPABE Department of Chemical Engineering, Faculty of Engineering, University of Porto, 4200-465 Porto, Portugal*

[4] *Instituto de Recursos Naturales y Agrobiología. IRNASA-CSIC, Salamanca, Spain.*

[5] *Unidad Asociada Universidad de Salamanca- CSIC 'Interacción Planta-Microorganismo'. Salamanca. Spain*

[6] *IPV - ESSV Polytechnic Institute of Viseu, Higher Health School of Viseu, 3500-843, Viseu, Portugal*

[7] *Departamento de Microbiología y Genética. Facultad de Farmacia. Universidad de Salamanca. Salamanca. Spain*

**Abstract:** Legumes are a wide group of plants worldwide cultivated by the high nutritional quality of their seeds containing fibre, proteins, resistant starch, minerals and vitamins. The milk obtained from seeds of some legumes, mainly soybean, is used to obtain fermented products currently widely consumed as probiotics. Moreover, legumes are considered as nutraceuticals since they contain bioactive peptides (BAPs) and many phytochemicals endowed with useful biological activities. Legume BAPs have antioxidant, antihypertensive, hypocholesterolemic and antithrombotic activities. Flavonoids and particularly isoflavones have beneficial effects in different cancer types, have been related with lower cardiovascular risk and are protective against fatty liver disease, obesity, diabetes and other metabolic disorders.

\* **Corresponding author Luís R. Silva:** CICS UBI Health Sciences Research Centre, University of Beira Interior, 6201-506 Covilhã, Portugal; Tel: +351 275 329 077; Fax: +351 275 329 099; Email: luisfarmacognosia@gmail.com.

**Luís Rodrigues da Silva and Branca Maria Silva (Eds.)**

Hydrophilic phytochemicals, such as ascorbic acid (vitamin C), phenolic acids and polyphenols, have been associated with a reduction of cancer risk and an enhancement of the immune system functionality. Lipophilic phytonutrients, such as carotenoids and tocopherols, may prevent the risk of cardiovascular diseases. In this chapter we revise the recent works focusing on legume bioactive compounds and human health prevention.

**Keywords:** Bioactive peptides, Carotenoids, Fatty acids, Isoflavones, Legumes, Phenolics, Tocopherols.

## INTRODUCTION

The term 'legumes' refers to a wide group of angiospermal plants worldwide distributed that are able to grow in diverse aquatic and terrestrial environments, under different edapho-climatic conditions. Legumes seeds (or pulses) constitute the main source of vegetal protein consumed in the world [1] as green or processed beans and as "milk", such as soy milk [2]. The most consumed legume worldwide is *Glycine max* (soy, soybean) which contains the highest quality protein found to date in plants [3] followed by *Arachis hypogaea* (peanut) and *Phaseolus vulgaris* (common bean), whose seeds are highly appreciated for their quality proteins [4 - 6].

In addition to the high nutritive quality of their proteins the comsumption of peas, beans and chickpea [7 - 10] has been related with health benefits. Some pulses such as *Lens culinaris* (lentil), *Vicia faba* (faba bean), *Pisum sativum* (pea) and *Cicer arietinum* (chickpea) are included in the Mediterranean diet, whose benefits for human health are well documented [8, 11 - 14]. Several reports showed that pulses are functional foods that combat obesity [15], reduce metabolic syndrome risk factors in overweight and obese adults [16] and prevent hypercholestero-lemia, hypertension, diabetes and cardiovascular and renal diseases [17 - 19].

Also, fermented derivatives of legumes, mainly those from soybean milk, are traditionally used worldwide as probiotics after fermentation with lactic bacteria, bifidobacteria and/or yeasts [20, 21]. Nevertheless in the last decade other pulses are explored as novel probiotics, such as peanut [22], lupin [23], pigeon pea [24], bambara groundnut [25], mung bean [26]. Even the fermentation of mixed legume milks such as those of peanut and soybean are being investigated [27]. Legumes also contain polysaccharides considered as prebiotic, such as the raffinose family of oligosaccharides present in lupin and soybean seeds [23, 28]. In lentil the polysaccharides with prebiotic potential include those from raffinose-family oligosaccharides, sugar alcohols, fructooligosaccharides, and resistant starch, which varies with the variety and the location [29] and it has been reported that pectic oligosaccharides derived from chickpea (*Cicer arietinum*) have prebiotic and antioxidant activities [30].

Legumes are part of the named nutraceutical products due to their benefits for human health mainly based on their bioactive compounds, including BAPs, phenolic compounds, carotenoids, tocopherols and fatty acids, among other phytochemicals [11, 31 - 37]. Most of legumes have been rarely studied to date, but the interest in the research about their benefits for human health is increasing and in the last years many reports about the nutritional characteristics, chemical composition and antioxidant potential of several underutilized legumes have been published [38 - 46].

**Bioactive Peptides (BAPs)**

The food proteins release peptides of variable size in the intestinal lumen, some of them resistant to further digestion, which in some cases share structural motifs along with endogenous peptides, for example endorphins or exorphins, known to modulate physiological functions [47 - 49]. BAPs from animal origin (milk, eggs, *etc.*) have been widely studied, but also from different legumes (soybean, pea, lentil, beans and chickpea) are used to obtain BAPs [50], being lunasin, from soybean, exploited as commercial source of BAPs [36].

BAPs are peptides encrypted in intact molecules, which are released by different enzymes during gastrointestinal transit or by fermentation or ripening of foods [47, 51]. It has been reported that they have positive effects for the human health such as immunomodulating, antihypertensive, osteoprotective, antilipemic, opiate-like, anti-thrombotic, antioxidative, anticariogenic and antimicrobial [47, 48, 51]. Although these effects have been mainly studied in milk proteins, pulses are also a rich protein source in human diet being their consumption associated with the prevention of different chronic diseases [36].

The most studied legume BAPs are those from soybean since soy milk and its fermented product are a good source of these peptides [52]. Lunasin, a heat stable soy-derived bioactive peptide with high bioavailability, is composed of 43 amino acid residues with a molecular weight of 5.5 kDa. Lunasin possesses anti-oxidative, anti-inflammatory and anticancerous properties having a role in the regulation of cholesterol biosynthesis [53]. Moreover, in a recent work it has been reported that legumin, the main seed protein of chickpea, was reported to be a source of antioxidant peptides with potential for developing new nutraceuticals and functional foods [54].

The BAPs of *P. vulgaris* (common bean), that is worldwide cultivated and considered a nutraceutical food and a good source of protein, have been recently studied in several works. Common bean hydrolysates and peptides have been reported to perform mainly angiotensin I converting enzyme inhibition (ACEI), antioxidant capacity, and antimicrobial and tumor cell inhibition activities [6]. For example BAPs of this legume have ACEI [55] and potential to reduce parameters related to the risk of developing type-2 diabetes (T2D) *in vitro* [56]. As occurs in the case of lunasin from soybean, the precooking of *P. vulgaris* (common bean) seeds does not affect the potential bioactivities of protein hydrolysates with pepsin/pancreatin, which include antioxidant activity and inhibition of α-amylase, α-glucosidase, dipeptidyl peptidase-IV (DPP-IV) and angiotensin converting enzyme I (ACE) [57].

In the last years the BAPs of other legumes are also being studied such as those from *Psophocarpus tetragonolobus* (winged bean) and *Lupinus* spp. (lupine) with high ACEI [58, 59], those from *Lens culinaris* (lentils) with antioxidant and ACEI activities [60], those from *Parkia speciosa* seeds (stinker bean) with antioxidative and antihypertensive activities [39], those from *Vigna unguiculata* (cowpea) with hypocholesterolemic activity [61], those from *Mucunapruriens* with ACEI, antioxidant and antithrombotic capacities [62] and those from *Cicer arietinum* (chickpea) with antioxidant activity [54, 63].

Since BAPs are released from food proteins in the human gut, microorganisms can play a crucial role in this process. The human gut microbiome is very complex and depending on its composition, the type of released peptides and their bioactivities can vary. For example, in a recent study it was reported that whereas *Bacillus subtilis* released peptides from kidney beans (*P. vulgaris*) with antioxidant activity, *Lactobacillus plantarum* released peptides with potential antihypertensive activity due to their high γ-aminobutyric acid (GABA) content and ACEI activity [64]. Therefore, the microbial metabolism modulates the type of BAPs released from legume proteins and different fermentation types influence the activity of protein lysates such as those of lentils *(Lens culinaris)* since water-soluble fractions obtained by liquid state fermentation had higher free amino groups, GABA content, antioxidant and ACEI activities than those obtained by solid state fermentation [60].

## Isoflavones

Isoflavones are a type of flavonoids present in legumes which together with phenolic acids and procyanidins constitute the major phenolic compounds [33, 65 - 67]. The benefits for human health of several phenolic compounds present in legume seeds have been analysed, nevertheless the most studied are the legume flavonoids, particularly soybean and red clover isoflavones, whose potential to protect against different diseases has been widely reported [68].

The role of isoflavones, particularly those from soybean, and in less extention from red clover [69], to prevent different pathologies has been widely studied, particularly cancer, menopausial symptoms, obesity and different metabolic diseases, and it has been suggested that optimal intakes of soya protein and isoflavones are 15-20g/d and 50-90mg/d, respectively [70]. The bioavailability of isoflavones after their intake is modulated by the gut microbiota and therefore the health effects of dietary phytoestrogens are strongly determined by the intestinal microbiota of each individual [71 - 73]. Absorption of isoflavones probably occurs in the small intestine which contains bacteria able to hydrolyze conjugated isoflavones, releasing the bioactive aglycones for direct absorption or further metabolism and reconjugation [72 - 74]. The effectiveness of soy isoflavones may depend on their biotransformation into equol, a metabolite with higher estrogenic activity, which may enhance the therapeutical activity of isoflavones [75, 76] and can regulate the cancer cell viability and the protein synthesis initiation *via* c-Myc and eIF4G [77]. Some authors have suggested that changing the intestinal microbiota to increase the proportion of equol-producing microorganisms could be a possible strategy for reducing the risk of prostatic cancer [78]. The lactic bacteria can convert isoflavone glucosides into aglycones during soymilk fermentation [79 - 81] and the complete genomes of some equol-producing lactic bacteria have been recently sequenced [82]. Also it has been reported that strains of *Bifidobacterium brevis* and *Bifidobacterium longum* are able to convert diadzein to equol [83]. Therefore currently many studies focus on the possibility to prevent different cancer types throuth different probiotic foods consumption [84] that can be fortified with the addition of physiologically active ingredients [85].

It has been suggested that through regulation of multiple pathways the isoflavone genistein is a potent inhibitor of angiogenesis, which is considered as a key step in cancer growth, invasion, and metastasis [86], and it decreases the stem-like cell population in breast cancer through Hedgehog pathway [87]. In addition to genistein, daidzein and resveratrol have also anticarcinogenic effects [88]. Since genistein is similar to estradiol-17beta molecules [88] isoflavones having an effect similar than the estrogens with the advantage of avoiding the risks linked to the estrogen treatments [89]. Since these treatments can alleviate "hot flush" in menopausial women [90, 91], it was assumed that soybean and red clover isoflavones have similar effects, but different studies reported contradictory results. Some authors concluded that isoflavones cannot be recommended for the relief of hot flushes [92], whereas others found reductions on hot flashes and co-occurring symptoms during the menopause [93 - 95]. Since the estrogen treatments also protect bones of postmenopausal women from osteoporosis reducing the fracture risk [96], the protection provided by isoflavonoids has also been analysed [97, 98]. A positive effect of isoflavone on bone mineral density was suggested after the revision of several studies [99], but other studies pointed out that the comsumption of soybean isoflavones had no added benefit in preventing bone loss [100].

The cancers related with estrogenic activity have been the most analysed as occurs with breast cancer [69, 84] with controversial results [101, 102]. Whereas some authors have been found positive effects [103], others found different responses in women from Asia and other countries after soybean isoflavone intake [104] and others observed a non statistically significant association between the intake of dietary isoflavones and breast cancer risk, although their results suggested possible ethnic/racial differences [105]. Some studies suggested a reduction of endometrial cancer risk in women after isoflavones intake [106, 107], whereas others do not found evidence of a protective association between soy food or isoflavone intake and endometrial cancer risk [108]. Nevertheless, a recent study in animal models showed that soybean isoflavones attenuate the expression of genes related to endometrial cancer risk [37] and some authors have found that soy and isoflavone intake are associated with a reduced ovarian cancer risk in Chinese women who consumed 120 g/day of soy foods [109]. Also a lower risk of prostate cancer was associated the consumption of soy foods [110, 111] and the higher breast and prostate cancer risk in Western countries than in Eastern ones has been attributed to a lower intake of soybean products [89]. Recently it has been reported that the isoflavone genistein inhibits human prostate cancer cell detachment, invasion, and metastasis [112].

The effects of isoflavones in other cancer types was studied as well since it has been reported that genistein inhibits human colorectal cancer growth [113] and it is able to induce apoptosis in colon cancer human lines [114, 115]. Several studies suggested that consumption of soy foods may reduce the colorectal cancer risk in women, but not in men [116 - 118]. Other studies related the soybean and isoflavone intake with a lower colorectal cancer risk in men and posmenopausal women, but not in premenopausal women [119] and other studies do not found effect of soybean intake in this type of cancer [120]. Given the contradictory results obtained in several studies Kocic *et al.* [121] pointed out that more studies are necessary to make public health recommendations. However, the results of a recent study can help to the interpretation of the previous results since they showed that isoflavones soy intake as a food group is only associated with a small reduction in colorectal cancer risk, but dietary isoflavone intakes suggested a stronger inverse association with this cancer [122].

In addition to these diseases, isoflavone intake has been related with lower cardiovascular risk [123 - 125] and some recent studies suggested that isoflavones reduce cholesterol, triglycerides and/or uric acid contents in serum [126 - 128], are protective against fatty liver disease [129] and type 2 diabetes [130, 131] and even they have been proposed to be part of anti-aging strategies [132].

## Other Phenolic Compounds

Phenolic compounds have a hydroxyl group (-OH) bonded directly to an aromatic hydrocarbon group showing high antioxidant activity [133, 134] which can be enhanced after fermentation with lactic bacteria [135]. Soybean seeds contain phenolic compounds in addition to isoflavones [136, 137], nevertheless, as expected, dark color (bronze, red and black) legume seeds, such as lentils, coloured beans, and black soybeans had significantly higher phenolic content and antioxidant capacity than those of pale colour (white, yellow and green) legumes, such as yellow soybeans and yellow and green peas [138]. Anthocyanins are natural pigments located in some legume seed coats [66] whose interest is increasing as safe food colorants with health-promoting properties [139] based on their antioxidant potential [140]. Anthocyanin consumption has been related with several benefits for human health, such as prevention of cardiovascular risk, cancer, obesity or diabetes [141] and they have anti-inflamatory activity [142]. Although there are no specific studies about the benefits of legume anthocyanins for human health, the antioxidant potential of these compounds has been explored in some legume seeds [138, 143, 144] and the usefulness as natural stain of these compounds has been investigated in some legume seeds, such as adzuki bean (*Vigna angularis*) [145].

Proanthocyanidins, also known as condensed tannins, constitute another class of phenolic substances having a positive role in human health particularly in prevention of cardiovascular and cancer diseases [146, 149]. Several legume seeds contain proanthocyanidins [67, 150] and extracts of lentils, red kidney beans, black soybeans and black beans were reported to be more effective in inhibiting LDL oxidation than yellow soybeans *in vitro* conditions [138]. Within proanthocyanidins the most important group are procyanidins composed exclusively of the monomeric flavan-3-ol-constituents epicatechin and catechin [67] that play important roles in human health [151, 152]. Different types of procyanidins have been found, for example, in lentil seeds [153, 154], adzuki bean seeds [145, 155].

## Carotenoids, Tocopherols and Fatty Acids

Legumes are a good source of antioxidants [66, 156] and their seeds contain carotenoids, tocopherols and fatty acids that are essential nutrients for humans, comprising vitamin precursors, antioxidants and essential omega-3 fatty acids [157]. The ten most economically important grain legumes, *Arachis* (peanut), *Cicer* (chickpea), *Glycine* (soybean), *Lathyrus* (vetch), *Lens* (lentil), *Lupinus* (lupin), *Phaseolus* (bean), *Pisum* (pea), *Vicia* (faba bean) and *Vigna* (cowpea) have different contents of carotenoids and tocopherols, being lutein the predominant carotenoid in all of them, followed by zeaxanthin and $\beta$-carotene [157]. Several studies showed that mainly soybean but also chickpea are rich in $\beta$-carotene [8, 157]. Carotenoids are natural plant pigments precursors of vitamin A that play a role in prevention of cancer, coronary diseases, age-related macular degeneration, cataract, and deficiency in carotenoids can cause blindness [158 - 161].

Tocopherols constitute a class of phytochemicals most of which having vitamin E activity. From all tocopherols, $\gamma$-tocopherol is the most abundant isoform in legumes such as soybean, chickpea, lentil, pea, common bean, broad bean and three lupin species [32, 157, 162], although in some cowpeas, peanuts, black-eyed and pinto beans, $\delta$-tocopherol and $\alpha$-tocopherol were the main isoforms [32, 157]. As was observed for carotenoids, soybean has the highest content in tocopherol among the ten legumes analysed by Fernández-Marín *et al.* [155]. Tocopherols plays crucial roles in anti-inflammatory processes [163, 164] and deficiency of this vitamin results in a range of disorders, including neuromuscular problems [165] and coronary diseases [162, 166]. Although the contribution of $\gamma$-tocopherol to the total vitamin E activity is only one tenth of that of $\alpha$-tocopherol, it has specific functions in the detoxification of nitrogen dioxide and other reactive nitrogen species. Some epidemiological studies suggested that $\gamma$-tocopherol may protect against cardiovascular diseases, because its plasma levels are inversely associated with increased morbidity and mortality [66].

Some legume seeds have great importance by their oil content, such as soybean and peanut and several studies focused on the fatty acid composition of several legume seeds [157, 167]. Some unsaturated fatty acids such as linoleic and oleic acids have been found in several legumes including soybean, chickpea, lentil, peanut and different types of beans and lupine [8, 32, 157, 167]. Linoleic acid is an essential omega-6 poly-unsaturated fatty acid that can improve insulin resistance reducing the incidence of diabetes being also associated the presence of higher levels of this fatty acid in blood with lower blood pressures [168]. Generally, a high intake of unsaturated fatty acids relative to saturated fatty acids prevents coronary disease [169] and omega-3 polyunsaturated fatty acids have anti-inflammatory action, and for this reason could be useful as an adjuvant in the treatment of some cancers [170].

## CONCLUDING REMARKS

All currently available studies showed the positive role of legumes in human health as functional foods or nutraceuticals due to their content in proteins, polysaccharides, vitamins, minerals, BAPs, flavonoids, carotenoids, tocopherols or fatty acids. Since some of these studies have been performed *in vitro* or in animal models further studies in humans are needed also extending the range of legumes analyzed since most of the edible legumes are not explored yet.

## CONFLICT OF INTEREST

The authors confirm that they have no conflict of interest to declare for this publication.

## ACKNOWLEDGEMENTS

This work was supported by the Portuguese "Fundação para a Ciência e a Tecnologia" - FCT: L.R. Silva (SFRH/BPD/105263/2014; CICS-UBI); UMIB (Pest-OE/SAU/UI0215/2014) co-funded by FEDER *via* Programa Operacional Fatores de Competitividade - COMPETE/QREN & FSE and POPH funds and Ministerio de Ciencia e Innovación (MINECO) and Junta de Castilla y León from Spain. Additionally, the authors would like to thank our numerous collaborators and students involved in this research over the years.

# REFERENCES

[1]     Tharanathan RN, Mahadevamma S. Grain legumes-a boon to human nutrition. Trends Food Sci Technol 2003; 14: 507-18.
[http://dx.doi.org/10.1016/j.tifs.2003.07.002]

[2]     Nishinari K, Fang Y, Guo S, Phillips GO. Soy proteins: A review on composition, aggregation and emulsification. Food Hydrocoll 2014; 39: 301-18.
[http://dx.doi.org/10.1016/j.foodhyd.2014.01.013]

[3]     Tome D. Criteria and markers for protein quality assessment - a review. Br J Nutr 2012; 108 (Suppl. 2): S222-9.
[http://dx.doi.org/10.1017/S0007114512002565] [PMID: 23107532]

[4]     Zhao X, Chen J, Du F. Potential use of peanut by-products in food processing: a review. J Food Sci Technol 2012; 49(5): 521-9.
[http://dx.doi.org/10.1007/s13197-011-0449-2] [PMID: 24082262]

[5]     Hayat I, Ahmad A, Masud T, Ahmed A, Bashir S. Nutritional and health perspectives of beans (Phaseolus vulgaris L.): an overview. Crit Rev Food Sci Nutr 2014; 54(5): 580-92.
[http://dx.doi.org/10.1080/10408398.2011.596639] [PMID: 24261533]

[6]     Luna-Vital DA, Mojica L, González de Mejía E, Mendoza S, Loarca-Piña G. Biological potential of protein hydrolysates and peptides from common bean (Phaseolus vulgaris L.): A review. Food Res Int 2015; 76: 39-50.
[http://dx.doi.org/10.1016/j.foodres.2014.11.024]

[7]     Dahl WJ, Foster LM, Tyler RT. Review of the health benefits of peas (*Pisum sativum* L.). Br J Nutr 2012; 108 (Suppl. 1): S3-S10.
[http://dx.doi.org/10.1017/S0007114512000852] [PMID: 22916813]

[8]     Jukanti AK, Gaur PM, Gowda CL, Chibbar RN. Nutritional quality and health benefits of chickpea (Cicer arietinum L.): a review. Br J Nutr 2012; 108 (Suppl. 1): S11-26.
[http://dx.doi.org/10.1017/S0007114512000797] [PMID: 22916806]

[9]     Hutchins AM, Winham DM, Thompson SV. Phaseolus beans: impact on glycaemic response and chronic disease risk in human subjects. Br J Nutr 2012; 108 (Suppl. 1): S52-65.
[http://dx.doi.org/10.1017/S0007114512000761] [PMID: 22916816]

[10]    Siah SD, Konczak I, Agboola S, Wood JA, Blanchard CL. *In vitro* investigations of the potential health benefits of Australian-grown faba beans (Vicia faba L.): chemopreventative capacity and inhibitory effects on the angiotensin-converting enzyme, α-glucosidase and lipase. Br J Nutr 2012; 108 (Suppl. 1): S123-34.
[http://dx.doi.org/10.1017/S0007114512000803] [PMID: 22916808]

[11]    Bouchenak M, Lamri-Senhadji M. Nutritional quality of legumes, and their role in cardiometabolic risk prevention: a review. J Med Food 2013; 16(3): 185-98.
[http://dx.doi.org/10.1089/jmf.2011.0238] [PMID: 23398387]

[12]    Castro-Quezada I, Román-Viñas B, Serra-Majem L. The Mediterranean diet and nutritional adequacy: a review. Nutrients 2014; 6(1): 231-48.
[http://dx.doi.org/10.3390/nu6010231] [PMID: 24394536]

[13] Sofi F, Macchi C, Abbate R, Gensini GF, Casini A. Mediterranean diet and health status: an updated meta-analysis and a proposal for a literature-based adherence score. Public Health Nutr 2014; 17(12): 2769-82.
[http://dx.doi.org/10.1017/S1368980013003169] [PMID: 24476641]

[14] Martínez-González MA, Salas-Salvadó J, Estruch R, Corella D, Fitó M, Ros E. Benefits of the Mediterranean Diet: Insights From the PREDIMED Study. Prog Cardiovasc Dis 2015; 58(1): 50-60.
[http://dx.doi.org/10.1016/j.pcad.2015.04.003] [PMID: 25940230]

[15] Marinangeli CP, Jones PJ. Pulse grain consumption and obesity: effects on energy expenditure, substrate oxidation, body composition, fat deposition and satiety. Br J Nutr 2012; 108 (Suppl. 1): S46-51.
[http://dx.doi.org/10.1017/S0007114512000773] [PMID: 22916815]

[16] Mollard RC, Luhovyy BL, Panahi S, Nunez M, Hanley A, Anderson GH. Regular consumption of pulses for 8 weeks reduces metabolic syndrome risk factors in overweight and obese adults. Br J Nutr 2012; 108 (Suppl. 1): S111-22.
[http://dx.doi.org/10.1017/S0007114512000712] [PMID: 22916807]

[17] Abeysekara S, Chilibeck PD, Vatanparast H, Zello GA. A pulse-based diet is effective for reducing total and LDL-cholesterol in older adults. Br J Nutr 2012; 108 (Suppl. 1): S103-10.
[http://dx.doi.org/10.1017/S0007114512000748] [PMID: 22916805]

[18] Singhal P, Kaushik G, Mathur P. Antidiabetic potential of commonly consumed legumes: a review. Crit Rev Food Sci Nutr 2014; 54(5): 655-72.
[http://dx.doi.org/10.1080/10408398.2011.604141] [PMID: 24261538]

[19] Arnoldi A, Zanoni C, Lammi C, Boschin G. The role of grain legumes in the prevention of hypercholesterolemia and hypertension. Crit Rev Plant Sci 2015; 34: 144-68.
[http://dx.doi.org/10.1080/07352689.2014.897908]

[20] Wang YC, Yu RC, Chou CC. Antioxidative activities of soymilk fermented with lactic acid bacteria and bifidobacteria. Food Microbiol 2006; 23(2): 128-35.
[http://dx.doi.org/10.1016/j.fm.2005.01.020] [PMID: 16942996]

[21] Rekha CR, Vijayalakshmi G. Biomolecules and nutritional quality of soymilk fermented with probiotic yeast and bacteria. Appl Biochem Biotechnol 2008; 151(2-3): 452-63.
[http://dx.doi.org/10.1007/s12010-008-8213-4] [PMID: 18607548]

[22] Bensmira M, Jiang B. Total phenolic compounds and antioxidant activity of a novel peanut based kefir. Food Sci Biotechnol 2015; 24: 1055-60.
[http://dx.doi.org/10.1007/s10068-015-0135-7]

[23] Martínez-Villaluenga C, Gómez R. Characterization of bifidobacteria as starters in fermented milk containing raffinose family of oligosaccharides from lupin as prebiotic. Int Dairy J 2007; 17: 116-22.
[http://dx.doi.org/10.1016/j.idairyj.2006.02.003]

[24] Parra K, Ferrer M, Piñero M, Barboza Y, Medina LM. Use of *Lactobacillus acidophilus* and Lactobacillus casei for a potential probiotic legume-based fermented product using pigeon pea (Cajanus cajan). J Food Prot 2013; 76(2): 265-71.
[http://dx.doi.org/10.4315/0362-028X.JFP-12-138] [PMID: 23433374]

[25]  Murevanhema YY, Jideani VA. Potential of Bambara groundnut (*Vigna subterranea* (L.) Verdc) milk as a probiotic beverage-a review. Crit Rev Food Sci Nutr 2013; 53(9): 954-67.
      [http://dx.doi.org/10.1080/10408398.2011.574803] [PMID: 23768187]

[26]  Wu H, Rui X, Li W, Chen X, Jiang M, Dong M. Mung bean (Vigna radiata) as probiotic food through fermentation with Lactobacillus plantarum B1-6. LWT -. Food Sci Technol (Campinas) 2015; 63: 445-51.

[27]  Santos CC, Libeck BdaS, Schwan RF. Co-culture fermentation of peanut-soy milk for the development of a novel functional beverage. Int J Food Microbiol 2014; 186: 32-41.
      [http://dx.doi.org/10.1016/j.ijfoodmicro.2014.06.011] [PMID: 24984220]

[28]  Wongputtisin P, Ramaraj R, Unpaprom Y, Kawaree R, Pongtrakul N. Raffinose family oligosaccharides in seed of Glycine max cv. Chiang Mai60 and potential source of prebiotic substances. Int J Food Sci Technol 2015; 50(8): 1750-6.
      [http://dx.doi.org/10.1111/ijfs.12842]

[29]  Johnson CS, Thavaraja D, Combs GF Jr, Thavarajah P. Lentil (*Lens culinaris* L.): A prebiotic-rich whole food legume. Food Res Int 2013; 51: 107-13.
      [http://dx.doi.org/10.1016/j.foodres.2012.11.025]

[30]  Shakuntala S, Mol P, Muralikrishna G. Pectic oligosaccharides derived from chickpea (*Cicer arietinum* L.) husk pectin and elucidation of their Role in prebiotic and antioxidant activities. Trends Carbohydr Res 2014; 6: 29-36.

[31]  Campos-Vega R, Loarca-Piña G, Dave Oomah B. Minor components of pulses and their potential impact on human health. Food Res Int 2010; 43: 461-82.
      [http://dx.doi.org/10.1016/j.foodres.2009.09.004]

[32]  Kalogeropoulos N, Chiou A, Ioannou M, Karathanos VT, Hassapidou M, Andrikopoulos NK. Nutritional evaluation and bioactive microconstituents (phytosterols, tocopherols, polyphenols, triterpenic acids) in cooked dry legumes usually consumed in the Mediterranean countries. Food Chem 2010; 121: 682-90.
      [http://dx.doi.org/10.1016/j.foodchem.2010.01.005]

[33]  Velázquez E, Silva LR, Peix A. Legumes: A healthy and ecological source of flavonoids. Curr Nutr Food Sci 2010; 6: 109-44.
      [http://dx.doi.org/10.2174/157340110791233247]

[34]  Wang S, Melnyk JP, Tsao R, Marcone MF. How natural dietary antioxidants in fruits, vegetables and legumes promote vascular health. Food Res Int 2011; 44: 14-22.
      [http://dx.doi.org/10.1016/j.foodres.2010.09.028]

[35]  Malaguti M, Dinelli G, Leoncini E, *et al.* Bioactive peptides in cereals and legumes: agronomical, biochemical and clinical aspects. Int J Mol Sci 2014; 15(11): 21120-35.
      [http://dx.doi.org/10.3390/ijms151121120] [PMID: 25405741]

[36]  López-Barrios L, Gutiérrez-Uribe JA, Serna-Saldívar SO. Bioactive peptides and hydrolysates from pulses and their potential use as functional ingredients. J Food Sci 2014; 79(3): R273-83.
      [http://dx.doi.org/10.1111/1750-3841.12365] [PMID: 24547749]

[37]  Carbonel AA, Calió ML, Santos MA, *et al.* Soybean isoflavones attenuate the expression of genes

related to endometrial cancer risk. Climacteric 2015; 18(3): 389-98.
[http://dx.doi.org/10.3109/13697137.2014.964671] [PMID: 25242508]

[38]    Annegowda HV, Bhat R, Tze LM, Karim AA, Mansor SM. The free radical scavenging and antioxidant activities of pod and seed extract of Clitoria fairchildiana (Howard)- an underutilized legume. J Food Sci Technol 2013; 50(3): 535-41.
[http://dx.doi.org/10.1007/s13197-011-0370-8] [PMID: 24425949]

[39]    Siow HL, Gan CY. Extraction of antioxidative and antihypertensive bioactive peptides from Parkia speciosa seeds. Food Chem 2013; 141(4): 3435-42.
[http://dx.doi.org/10.1016/j.foodchem.2013.06.030] [PMID: 23993504]

[40]    Katoch R. Nutritional potential of rice bean (*Vigna umbellata*): an underutilized legume. J Food Sci 2013; 78(1): C8-C16.
[http://dx.doi.org/10.1111/j.1750-3841.2012.02989.x] [PMID: 23278402]

[41]    Marimuthu M, Krishnamoorthi K. Nutrients and functional properties of horse gram (*Macrotyloma uniflorum*), an underutilized south Indian food legume. J Chem Pharm Res 2013; 5: 390-4.

[42]    Arivalagan M, Prasad TV, Harinder S, Ashok K. Variability in biochemical and mineral composition of *Mucuna pruriens* (L.) DC. – an underutilized tropical legume. Legume Res 2014; 37: 483-91.
[http://dx.doi.org/10.5958/0976-0571.2014.00664.X]

[43]    Leite Tavares R, Silva AS, Nascimento Campos AN, Pereira Schuler AR, de Souza Aquino S. Nutritional composition, phytochemicals and microbiological quality of the legume, *Mucuna pruriens.* Afr J Biotechnol 2015; 14: 676-82.
[http://dx.doi.org/10.5897/AJB2014.14354]

[44]    Nyau V, Prakash S, Rodrigues J, Farrant J. Antioxidant activities of bambara groundnuts as assessed by FRAP and DPPH assays. Am J Food Nutr 2015; 3: 7-11.

[45]    Prasad SK, Singh MK. Horse gram- an underutilized nutraceutical pulse crop: a review. J Food Sci Technol 2015; 52(5): 2489-99.
[http://dx.doi.org/10.1007/s13197-014-1312-z] [PMID: 25892749]

[46]    Sathya A, Siddhuraju P. Effect of processing methods on compositional evaluation of underutilized legume, Parkia roxburghii G. Don (yongchak) seeds. J Food Sci Technol 2015; 52(10): 6157-69.
[http://dx.doi.org/110.1007/s13197-015-1732-4]

[47]    Moughan PJ, Rutherfurd SM, Montoya CA, Dave LA. Food-derived bioactive peptides--a new paradigm. Nutr Res Rev 2014; 27(1): 16-20.
[http://dx.doi.org/10.1017/S0954422413000206] [PMID: 24231033]

[48]    Udenigwe CC, Aluko RE. Food protein-derived bioactive peptides: production, processing, and potential health benefits. J Food Sci 2012; 77(1): R11-24.
[http://dx.doi.org/10.1111/j.1750-3841.2011.02455.x] [PMID: 22260122]

[49]    Carbonaro M, Maselli P, Nucara A. Structural aspects of legume proteins and nutraceutical properties. Food Res Int 2015; 76: 19-30.
[http://dx.doi.org/10.1016/j.foodres.2014.11.007]

[50]    Roy F, Boye JI, Simpson BK. Bioactive proteins and peptides in pulse crops: Pea, chickpea and lentil. Food Res Int 2010; 43: 432-42.

[http://dx.doi.org/10.1016/j.foodres.2009.09.002]

[51]   Möller NP, Scholz-Ahrens KE, Roos N, Schrezenmeir J. Bioactive peptides and proteins from foods: indication for health effects. Eur J Nutr 2008; 47(4): 171-82.
[http://dx.doi.org/10.1007/s00394-008-0710-2] [PMID: 18506385]

[52]   Singh BP, Vij S, Hati S. Functional significance of bioactive peptides derived from soybean. Peptides 2014; 54: 171-9.
[http://dx.doi.org/10.1016/j.peptides.2014.01.022] [PMID: 24508378]

[53]   Lule VK, Garg S, Pophaly SD, Hitesh , Tomar SK. "Potential health benefits of lunasin: a multifaceted soy-derived bioactive peptide". J Food Sci 2015; 80(3): R485-94.
[http://dx.doi.org/10.1111/1750-3841.12786] [PMID: 25627564]

[54]   Torres-Fuentes C, Contreras MdelM, Recio I, Alaiz M, Vioque J. Identification and characterization of antioxidant peptides from chickpea protein hydrolysates. Food Chem 2015; 180: 194-202.
[http://dx.doi.org/10.1016/j.foodchem.2015.02.046] [PMID: 25766818]

[55]   Rui X, Boyea JI, Simpson BK, Prasher SO. Purification and characterization of angiotensin I-converting enzyme inhibitory peptides of small red bean (*Phaseolus vulgaris*) hydrolysates. J Funct Foods 2013; 5: 1116-24.
[http://dx.doi.org/10.1016/j.jff.2013.03.008]

[56]   de Souza Rocha T, Real Hernandez LM, Mojica L, Johnson MH, Chang YK, González de Mejía E. Germination of Phaseolus vulgaris and alcalase hydrolysis of its proteins produced bioactive peptides capable of improving markers related to type-2 diabetes *in vitro*. Food Res Int 2015; 76: 150-9.
[http://dx.doi.org/10.1016/j.foodres.2015.04.041]

[57]   Mojica L, Chen K, de Mejía EG. Impact of commercial precooking of common bean (*Phaseolus vulgaris*) on the generation of peptides, after pepsin-pancreatin hydrolysis, capable to inhibit dipeptidyl peptidase-IV. J Food Sci 2015; 80(1): H188-98.
[http://dx.doi.org/10.1111/1750-3841.12726] [PMID: 25495131]

[58]   Boschin G, Scigliuolo GM, Resta D, Arnoldi A. ACE-inhibitory activity of enzymatic protein hydrolysates from lupin and other legumes. Food Chem 2014; 145: 34-40.
[http://dx.doi.org/10.1016/j.foodchem.2013.07.076] [PMID: 24128446]

[59]   Wan Mohtar WA, Hamid AA, Abd-Aziz S, Muhamad SK, Saari N. Preparation of bioactive peptides with high angiotensin converting enzyme inhibitory activity from winged bean [*Psophocarpus tetragonolobus* (L.) DC.] seed. J Food Sci Technol 2014; 51(12): 3658-68.
[http://dx.doi.org/10.1007/s13197-012-0919-1] [PMID: 25477632]

[60]   Torino MI, Limón RI, Martínez-Villaluenga C, *et al.* Antioxidant and antihypertensive properties of liquid and solid state fermented lentils. Food Chem 2013; 136(2): 1030-7.
[http://dx.doi.org/10.1016/j.foodchem.2012.09.015] [PMID: 23122159]

[61]   Rodrigues Marques M, Fontanari G, Carvalho Pimenta D, Manólio Soares-Freitas R, Gomes Arêas JA. Proteolytic hydrolysis of cowpea proteins is able to release peptides with hypocholesterolemic activity. Food Res Int 2015; 77: 43-8.
[http://dx.doi.org/10.1016/j.foodres.2015.04.020]

[62]   Segura-Campos MR, Tovar-Benítez T, Chel-Guerrero L, Betancur-Ancona D. Functional and

bioactive properties of Velvet bean (Mucuna pruriens) protein hydrolysates produced by enzymatic treatments. J Food Measure Character 2014; 8: 61-9.
[http://dx.doi.org/10.1007/s11694-013-9165-0]

[63]   Kou X, Gao J, Xue Z, Zhang Z, Wang H, Wang X. Purification and identification of antioxidant peptides from chickpea (*Cicer arietinum* L.) albumin hydrolysates. LWT -. Food Sci Technol (Campinas) 2013; 50: 591-8.

[64]   Limón RI, Peñas E, Torino MI, Martínez-Villaluenga C, Dueñas M, Frias J. Fermentation enhances the content of bioactive compounds in kidney bean extracts. Food Chem 2015; 172: 343-52.
[http://dx.doi.org/10.1016/j.foodchem.2014.09.084] [PMID: 25442563]

[65]   Lin PY, Lai HM. Bioactive compounds in legumes and their germinated products. J Agric Food Chem 2006; 54(11): 3807-14.
[http://dx.doi.org/10.1021/jf060002o] [PMID: 16719500]

[66]   Amarowicz R, Pegg RB. Legumes as a source of natural antioxidants. Eur J Lipid Sci Technol 2008; 110: 865-78.
[http://dx.doi.org/10.1002/ejlt.200800114]

[67]   Bittner K, Rzeppa S, Humpf HU. Distribution and quantification of flavan-3-ols and procyanidins with low degree of polymerization in nuts, cereals, and legumes. J Agric Food Chem 2013; 61(38): 9148-54.
[http://dx.doi.org/10.1021/jf4024728] [PMID: 23971434]

[68]   Messina M. A brief historical overview of the past two decades of soy and isoflavone research. J Nutr 2010; 140(7): 1350S-4S.
[http://dx.doi.org/10.3945/jn.109.118315] [PMID: 20484551]

[69]   Fritz H, Seely D, Flower G, *et al.* Soy, red clover, and isoflavones and breast cancer: a systematic review. PLoS One 2013; 8(11): e81968.
[http://dx.doi.org/10.1371/journal.pone.0081968] [PMID: 24312387]

[70]   Messina M. Investigating the optimal soy protein and isoflavone intakes for women: a perspective. Womens Health (Lond Engl) 2008; 4(4): 337-56.
[http://dx.doi.org/10.2217/17455057.4.4.337] [PMID: 19072500]

[71]   Franke AA, Lai JF, Halm BM. Absorption, distribution, metabolism, and excretion of isoflavonoids after soy intake. Arch Biochem Biophys 2014; 559: 24-8.
[http://dx.doi.org/10.1016/j.abb.2014.06.007] [PMID: 24946051]

[72]   Landete JM, Curiel JA, Rodríguez H, de las Rivas B, Muñoz R. Aryl glycosidases from *Lactobacillus plantarum* increase antioxidant activity of phenolic compounds. J Funct Foods 2014; 7: 322-9.
[http://dx.doi.org/10.1016/j.jff.2014.01.028]

[73]   Marín L, Miguélez EM, Villar CJ, Lombó F. Bioavailability of dietary polyphenols and gut microbiota metabolism: antimicrobial properties. Biomed Res Int 2015; 2015: 905215.
[http://dx.doi.org/10.1155/2015/905215]

[74]   Vitale DC, Piazza C, Melilli B, Drago F, Salomone S. Isoflavones: estrogenic activity, biological effect and bioavailability. Eur J Drug Metab Pharmacokinet 2013; 38(1): 15-25.
[http://dx.doi.org/10.1007/s13318-012-0112-y] [PMID: 23161396]

[75]   Shor D, Sathyapalan T, Atkin SL, Thatcher NJ. Does equol production determine soy endocrine effects? Eur J Nutr 2012; 51(4): 389-98.
[http://dx.doi.org/10.1007/s00394-012-0331-7] [PMID: 22366740]

[76]   Rafii F. The role of colonic bacteria in the metabolism of the natural isoflavone daidzin to equol. Metabolites 2015; 5(1): 56-73.
[http://dx.doi.org/10.3390/metabo5010056] [PMID: 25594250]

[77]   de la Parra C, Borrero-Garcia LD, Cruz-Collazo A, Schneider RJ, Dharmawardhane S. Equol, an isoflavone metabolite, regulates cancer cell viability and protein synthesis initiation *via* c-Myc and eIF4G. J Biol Chem 2015; 290(10): 6047-57.
[http://dx.doi.org/10.1074/jbc.M114.617415] [PMID: 25593313]

[78]   Sugiyama Y, Masumori N, Fukuta F, *et al.* Influence of isoflavone intake and equol-producing intestinal flora on prostate cancer risk. Asian Pac J Cancer Prev 2013; 14(1): 1-4.
[http://dx.doi.org/10.7314/APJCP.2013.14.1.1] [PMID: 23534704]

[79]   Di Cagno R, Mazzacane F, Rizzello CG, *et al.* Synthesis of isoflavone aglycones and equol in soy milks fermented by food-related lactic acid bacteria and their effect on human intestinal Caco-2 cells. J Agric Food Chem 2010; 58(19): 10338-46.
[http://dx.doi.org/10.1021/jf101513r] [PMID: 20822177]

[80]   Shimada Y, Takahashi M, Miyazawa N, Abiru Y, Uchiyama S, Hishigaki H. Identification of a novel dihydrodaidzein racemase essential for biosynthesis of equol from daidzein in *Lactococcus* sp. strain 20-92. Appl Environ Microbiol 2012; 78(14): 4902-7.
[http://dx.doi.org/10.1128/AEM.00410-12] [PMID: 22582059]

[81]   Hati S, Vij S, Singh BP, Mandal S. β-Glucosidase activity and bioconversion of isoflavones during fermentation of soymilk. J Sci Food Agric 2015; 95(1): 216-20.
[http://dx.doi.org/10.1002/jsfa.6743] [PMID: 24838442]

[82]   Toh H, Oshima K, Suzuki T, Hattori M, Morita H. Complete genome sequence of the equol-producing bacterium *Adlercreutzia equolifaciens* DSM 19450T. Genome Announc 2013; 1(5): e00742-13.
[http://dx.doi.org/10.1128/genomeA.00742-13] [PMID: 24051320]

[83]   Elghali S, Mustafa S, Amid M, Manap MY, Ismail A, Abas F. Bioconversion of daidzein to equol by Bifidobacterium breve 15700 and *Bifidobacterium longum* BB536. J Funct Foods 2012; 4: 736-45.
[http://dx.doi.org/10.1016/j.jff.2012.04.013]

[84]   Takagi A, Kano M, Kaga C. Possibility of breast cancer prevention: use of soy isoflavones and fermented soy beverage produced using probiotics. Int J Mol Sci 2015; 16(5): 10907-20.
[http://dx.doi.org/10.3390/ijms160510907] [PMID: 25984609]

[85]   Fayed AE. Review article: health benefits of some physiologically active ingredients and their suitability as yoghurt fortifiers. J Food Sci Technol 2015; 52(5): 2512-21.
[http://dx.doi.org/10.1007/s13197-014-1393-8] [PMID: 25892751]

[86]   Varinska L, Gal P, Mojzisova G, Mirossay L, Mojzis J. Soy and breast cancer: focus on angiogenesis. Int J Mol Sci 2015; 16(5): 11728-49.
[http://dx.doi.org/10.3390/ijms160511728] [PMID: 26006245]

[87]   Fan P, Fan S, Wang H, *et al.* Genistein decreases the breast cancer stem-like cell population through

Hedgehog pathway. Stem Cell Res Ther 2013; 4(6): 146.
[http://dx.doi.org/10.1186/scrt357] [PMID: 24331293]

[88]   Hwang KA, Choi KC. Anticarcinogenic effects of dietary phytoestrogens and their chemopreventive mechanisms. Nutr Cancer 2015; 67(5): 796-803.
[http://dx.doi.org/10.1080/01635581.2015.1040516] [PMID: 25996655]

[89]   Vaya J, Tamir S. The relation between the chemical structure of flavonoids and their estrogen-like activities. Curr Med Chem 2004; 11(10): 1333-43.
[http://dx.doi.org/10.2174/0929867043365251] [PMID: 15134523]

[90]   Andrikoula M, Prelevic G. Menopausal hot flushes revisited. Climacteric 2009; 12(1): 3-15.
[http://dx.doi.org/10.1080/13697130802556296] [PMID: 19061056]

[91]   Messina M. Soy foods, isoflavones, and the health of postmenopausal women. Am J Clin Nutr 2014; 100 (Suppl. 1): 423S-30S.
[http://dx.doi.org/10.3945/ajcn.113.071464] [PMID: 24898224]

[92]   Eden JA. Phytoestrogens for menopausal symptoms: a review. Maturitas 2012; 72(2): 157-9.
[http://dx.doi.org/10.1016/j.maturitas.2012.03.006] [PMID: 22516278]

[93]   Aso T, Uchiyama S, Matsumura Y, *et al.* A natural S-equol supplement alleviates hot flushes and other menopausal symptoms in equol nonproducing postmenopausal Japanese women. J Womens Health (Larchmt) 2012; 21(1): 92-100.
[http://dx.doi.org/10.1089/jwh.2011.2753] [PMID: 21992596]

[94]   Messina M. Soybean isoflavones warrant greater consideration as a treatment for the alleviation of menopausal hot flashes. Womens Health (Lond Engl) 2014; 10(6): 549-53.
[http://dx.doi.org/10.2217/whe.14.38] [PMID: 25482479]

[95]   Thomas AJ, Ismail R, Taylor-Swanson L, *et al.* Effects of isoflavones and amino acid therapies for hot flashes and co-occurring symptoms during the menopausal transition and early postmenopause: a systematic review. Maturitas 2014; 78(4): 263-76.
[http://dx.doi.org/10.1016/j.maturitas.2014.05.007] [PMID: 24951101]

[96]   Cauley JA. Estrogen and bone health in men and women. Steroids 2015; 99(Pt A): 11-5.
[http://dx.doi.org/10.1016/j.steroids.2014.12.010] [PMID: 25555470]

[97]   Ishimi Y. Soybean isoflavones in bone health. Forum Nutr 2009; 61: 104-16.
[http://dx.doi.org/10.1159/000212743] [PMID: 19367115]

[98]   Chin KY, Ima-Nirwana S. Can soy prevent male osteoporosis? A review of the current evidence. Curr Drug Targets 2013; 14(14): 1632-41.
[http://dx.doi.org/10.2174/13894501146661311216222612] [PMID: 24354587]

[99]   Castelo-Branco C, Soveral I. Phytoestrogens and bone health at different reproductive stages. Gynecol Endocrinol 2013; 29(8): 735-43.
[http://dx.doi.org/10.3109/09513590.2013.801441] [PMID: 23741966]

[100]  D'Adamo CR, Sahin A. Soy foods and supplementation: a review of commonly perceived health benefits and risks. Altern Ther Health Med 2014; 20 (Suppl. 1): 39-51.
[PMID: 24473985]

[101]  Andres S, Abraham K, Appel KE, Lampen A. Risks and benefits of dietary isoflavones for cancer. Crit Rev Toxicol 2011; 41(6): 463-506.
[http://dx.doi.org/10.3109/10408444.2010.541900] [PMID: 21438720]

[102]  Kwon Y. Effect of soy isoflavones on the growth of human breast tumors: findings from preclinical studies. Food Sci Nutr 2014; 2(6): 613-22.
[http://dx.doi.org/10.1002/fsn3.142] [PMID: 25493176]

[103]  Nagata C, Mizoue T, Tanaka K, *et al*. Soy intake and breast cancer risk: an evaluation based on a systematic review of epidemiologic evidence among the Japanese population. Jpn J Clin Oncol 2014; 44(3): 282-95.
[http://dx.doi.org/10.1093/jjco/hyt203] [PMID: 24453272]

[104]  Chen M, Rao Y, Zheng Y, *et al*. Association between soy isoflavone intake and breast cancer risk for pre- and post-menopausal women: a meta-analysis of epidemiological studies. PLoS One 2014; 9(2): e89288.
[http://dx.doi.org/10.1371/journal.pone.0089288] [PMID: 24586662]

[105]  Morimoto Y, Maskarinec G, Park SY, *et al*. Dietary isoflavone intake is not statistically significantly associated with breast cancer risk in the Multiethnic Cohort. Br J Nutr 2014; 112(6): 976-83.
[http://dx.doi.org/10.1017/S0007114514001780] [PMID: 25201305]

[106]  Bandera EV, Williams MG, Sima C, *et al*. Phytoestrogen consumption and endometrial cancer risk: a population-based case-control study in New Jersey. Cancer Causes Control 2009; 20(7): 1117-27.
[http://dx.doi.org/10.1007/s10552-009-9336-9] [PMID: 19353280]

[107]  Ollberding NJ, Lim U, Wilkens LR, *et al*. Legume, soy, tofu, and isoflavone intake and endometrial cancer risk in postmenopausal women in the multiethnic cohort study. J Natl Cancer Inst 2012; 104(1): 67-76.
[http://dx.doi.org/10.1093/jnci/djr475] [PMID: 22158125]

[108]  Budhathoki S, Iwasaki M, Sawada N, *et al*. Soy food and isoflavone intake and endometrial cancer risk: the Japan Public Health Center-based prospective study. BJOG 2015; 122(3): 304-11.
[http://dx.doi.org/10.1111/1471-0528.12853] [PMID: 24941880]

[109]  Lee AH, Su D, Pasalich M, Tang L, Binns CW, Qiu L. Soy and isoflavone intake associated with reduced risk of ovarian cancer in southern Chinese women. Nutr Res 2014; 34(4): 302-7.
[http://dx.doi.org/10.1016/j.nutres.2014.02.005] [PMID: 24774066]

[110]  Yan L, Spitznagel EL. Soy consumption and prostate cancer risk in men: a revisit of a meta-analysis. Am J Clin Nutr 2009; 89(4): 1155-63.
[http://dx.doi.org/10.3945/ajcn.2008.27029] [PMID: 19211820]

[111]  van Die MD, Bone KM, Williams SG, Pirotta MV. Soy and soy isoflavones in prostate cancer: a systematic review and meta-analysis of randomized controlled trials. BJU Int 2014; 113(5b): E119-30.
[http://dx.doi.org/10.1111/bju.12435] [PMID: 24053483]

[112]  Pavese JM, Krishna SN, Bergan RC, Takagi A, Kano M, Kaga C. Genistein inhibits human prostate cancer cell detachment, invasion, and metastasis. Am J Clin Nutr 2014; 100 (Suppl. 1): 431S-6S.
[http://dx.doi.org/10.3945/ajcn.113.071290] [PMID: 24871471]

[113]  Qin J, Chen JX, Zhu Z, Teng JA. Genistein inhibits human colorectal cancer growth and suppresses

miR-95, Akt and SGK1. Cell Physiol Biochem 2015; 35(5): 2069-77.
[http://dx.doi.org/10.1159/000374013] [PMID: 25871428]

[114]   Zhang Z, Wang CZ, Du GJ, *et al.* Genistein induces G2/M cell cycle arrest and apoptosis *via* ATM/p53-dependent pathway in human colon cancer cells. Int J Oncol 2013; 43(1): 289-96.
[PMID: 23686257]

[115]   Luo Y, Wang SX, Zhou ZQ, *et al.* Apoptotic effect of genistein on human colon cancer cells *via* inhibiting the nuclear factor-kappa B (NF-κB) pathway. Tumour Biol 2014; 35(11): 11483-8.
[http://dx.doi.org/10.1007/s13277-014-2487-7] [PMID: 25128065]

[116]   Oba S, Nagata C, Shimizu N, *et al.* Soy product consumption and the risk of colon cancer: a prospective study in Takayama, Japan. Nutr Cancer 2007; 57(2): 151-7.
[http://dx.doi.org/10.1080/01635580701274475] [PMID: 17571948]

[117]   Yang G, Shu XO, Li H, *et al.* Prospective cohort study of soy food intake and colorectal cancer risk in women. Am J Clin Nutr 2009; 89(2): 577-83.
[http://dx.doi.org/10.3945/ajcn.2008.26742] [PMID: 19073792]

[118]   Yan L, Spitznagel EL, Bosland MC. Soy consumption and colorectal cancer risk in humans: a meta-analysis. Cancer Epidemiol Biomarkers Prev 2010; 19(1): 148-58.
[http://dx.doi.org/10.1158/1055-9965.EPI-09-0856] [PMID: 20056634]

[119]   Budhathoki S, Joshi AM, Ohnaka K, *et al.* Soy food and isoflavone intake and colorectal cancer risk: the Fukuoka Colorectal Cancer Study. Scand J Gastroenterol 2011; 46(2): 165-72.
[http://dx.doi.org/10.3109/00365521.2010.522720] [PMID: 20969489]

[120]   Akhter M, Inoue M, Kurahashi N, Iwasaki M, Sasazuki S, Tsugane S. Dietary soy and isoflavone intake and risk of colorectal cancer in the Japan public health center-based prospective study. Cancer Epidemiol Biomarkers Prev 2008; 17(8): 2128-35.
[http://dx.doi.org/10.1158/1055-9965.EPI-08-0182] [PMID: 18708407]

[121]   Kocic B, Kitic D, Brankovic S. Dietary flavonoid intake and colorectal cancer risk: evidence from human population studies. J BUON 2013; 18(1): 34-43.
[PMID: 23613386]

[122]   Tse G, Eslick GD. Soy and isoflavone consumption and risk of gastrointestinal cancer: a systematic review and meta-analysis. Eur J Nutr 2014; •••
[http://dx.doi.org/10.1007/s00394-014-0824-7] [PMID: 25547973]

[123]   McCullough ML, Peterson JJ, Patel R, Jacques PF, Shah R, Dwyer JT. Flavonoid intake and cardiovascular disease mortality in a prospective cohort of US adults. Am J Clin Nutr 2012; 95(2): 454-64.
[http://dx.doi.org/10.3945/ajcn.111.016634] [PMID: 22218162]

[124]   Rebholz CM, Reynolds K, Wofford MR, *et al.* Effect of soybean protein on novel cardiovascular disease risk factors: a randomized controlled trial. Eur J Clin Nutr 2013; 67(1): 58-63.
[http://dx.doi.org/10.1038/ejcn.2012.186] [PMID: 23187956]

[125]   Ponzo V, Goitre I, Fadda M, *et al.* Dietary flavonoid intake and cardiovascular risk: a population-based cohort study. J Transl Med 2015; 13: 218.
[http://dx.doi.org/10.1186/s12967-015-0573-2] [PMID: 26152229]

[126]  Wong JM, Kendall CW, Marchie A, *et al.* Equol status and blood lipid profile in hyperlipidemia after consumption of diets containing soy foods. Am J Clin Nutr 2012; 95(3): 564-71.
[http://dx.doi.org/10.3945/ajcn.111.017418] [PMID: 22301925]

[127]  Benkhedda K, Boudrault C, Sinclair S, Marles R, Xiao C. A systematic review and meta-analysis of the effects of soy products on blood cholesterol levels. FASEB J 2015; 29: 923-14.

[128]  Qin Y, Shu F, Zeng Y, *et al.* Daidzein supplementation decreases serum triglyceride and uric acid concentrations in hypercholesterolemic adults with the effect on triglycerides being greater in those with the GA compared with the GG genotype of ESR-β RsaI. J Nutr 2014; 144(1): 49-54.
[http://dx.doi.org/10.3945/jn.113.182725] [PMID: 24225450]

[129]  Qiu LX, Chen T. Novel insights into the mechanisms whereby isoflavones protect against fatty liver disease. World J Gastroenterol 2015; 21(4): 1099-107.
[http://dx.doi.org/10.3748/wjg.v21.i4.1099] [PMID: 25632182]

[130]  Behloul N, Wu G. Genistein: a promising therapeutic agent for obesity and diabetes treatment. Eur J Pharmacol 2013; 698(1-3): 31-8.
[http://dx.doi.org/10.1016/j.ejphar.2012.11.013] [PMID: 23178528]

[131]  Talaei M, Pan A. Role of phytoestrogens in prevention and management of type 2 diabetes. World J Diabetes 2015; 6(2): 271-83.
[http://dx.doi.org/10.4239/wjd.v6.i2.271] [PMID: 25789108]

[132]  Park S, Yang MJ, Ha SN, Lee JS. Effective Anti-aging Strategies in an Era of Super-aging. J Menopausal Med 2014; 20(3): 85-9.
[http://dx.doi.org/10.6118/jmm.2014.20.3.85] [PMID: 25580418]

[133]  Djordjevic TM, Šiler-Marinkovic SS, Dimitrijevic-Brankovic SI. Antioxidant activity and total phenolic content in some cereals and legumes. Int J Food Proper 2011; 14: 175-84.
[http://dx.doi.org/10.1080/10942910903160364]

[134]  Bubols GB, Vianna DdaR, Medina-Remon A, *et al.* The antioxidant activity of coumarins and flavonoids. Mini Rev Med Chem 2013; 13(3): 318-34.
[PMID: 22876957]

[135]  Landete JM, Arqués J, Medina M, Gaya P, De La Rivas B, Muñoz R. Bioactivation of Phytoestrogens: Intestinal Bacteria and Health. Crit Rev Food Sci Nutr 2015. [Epub ahead of print]
[http://dx.doi.org/10.1080/10408398.2013.789823] [PMID: 25848676]

[136]  Tyug TS, Prasad NK, Ismail A. Antioxidant capacity, phenolics and isoflavones in soybean by-products. Food Chem 2010; 123: 583-9.
[http://dx.doi.org/10.1016/j.foodchem.2010.04.074]

[137]  Kim EH, Lee OK, Kim JK, *et al.* Isoflavones and anthocyanins analysis in soybean (Glycine max (L.) Merill) from three different planting locations in Korea. Field Crops Res 2014; 156: 76-83.
[http://dx.doi.org/10.1016/j.fcr.2013.10.020]

[138]  Xu BJ, Yuan SH, Chang SK. Comparative analyses of phenolic composition, antioxidant capacity, and color of cool season legumes and other selected food legumes. J Food Sci 2007; 72(2): S167-77.
[http://dx.doi.org/10.1111/j.1750-3841.2006.00261.x] [PMID: 17995859]

[139] He J, Giusti MM. Anthocyanins: natural colorants with health-promoting properties. Annu Rev Food Sci Technol 2010; 1: 163-87.
[http://dx.doi.org/10.1146/annurev.food.080708.100754] [PMID: 22129334]

[140] Leopoldini M, Russo N, Toscano M. The molecular basis of working mechanism of natural polyphenolic antioxidants. Food Chem 2011; 125: 288-306.
[http://dx.doi.org/10.1016/j.foodchem.2010.08.012]

[141] Pojer E, Mattivi F, Johnson D, Stockley CS. The case for anthocyanin consumption to promote human health: a review. Compr Rev Food Sci Food Saf 2013; 12: 483-508.
[http://dx.doi.org/10.1111/1541-4337.12024]

[142] Vendrame S, Klimis-Zacas D. Anti-inflammatory effect of anthocyanins *via* modulation of nuclear factor-κB and mitogen-activated protein kinase signaling cascades. Nutr Rev 2015; 73(6): 348-58.
[http://dx.doi.org/10.1093/nutrit/nuu066] [PMID: 26011910]

[143] Xu B, Chang SK. Antioxidant capacity of seed coat, dehulled bean, and whole black soybeans in relation to their distributions of total phenolics, phenolic acids, anthocyanins, and isoflavones. J Agric Food Chem 2008; 56(18): 8365-73.
[http://dx.doi.org/10.1021/jf801196d] [PMID: 18729453]

[144] Golam Masum Akond AS, Khandaker L, Berthold J, *et al.* Anthocyanin, total polyphenols and antioxidant activity of common bean. Am J Food Technol 2011; 6: 385-94.
[http://dx.doi.org/10.3923/ajft.2011.385.394]

[145] Han KH, Kitano-Okada T, Seo JM, *et al.* Characterisation of anthocyanins and proanthocyanidins of adzuki bean extracts and their antioxidant activity. J Funct Foods 2015; 14: 692-701.
[http://dx.doi.org/10.1016/j.jff.2015.02.018]

[146] Aron PM, Kennedy JA. Flavan-3-ols: nature, occurrence and biological activity. Mol Nutr Food Res 2008; 52(1): 79-104.
[http://dx.doi.org/10.1002/mnfr.200700137] [PMID: 18081206]

[147] Nandakumar V, Singh T, Katiyar SK. Multi-targeted prevention and therapy of cancer by proanthocyanidins. Cancer Lett 2008; 269(2): 378-87.
[http://dx.doi.org/10.1016/j.canlet.2008.03.049] [PMID: 18457915]

[148] Schroeter H, Heiss C, Spencer JP, Keen CL, Lupton JR, Schmitz HH. Recommending flavanols and procyanidins for cardiovascular health: current knowledge and future needs. Mol Aspects Med 2010; 31(6): 546-57.
[http://dx.doi.org/10.1016/j.mam.2010.09.008] [PMID: 20854838]

[149] Osakabe N. Flavan 3-ols improve metabolic syndrome risk factors: evidence and mechanisms. J Clin Biochem Nutr 2013; 52(3): 186-92.
[http://dx.doi.org/10.3164/jcbn.12-130] [PMID: 23704807]

[150] Jin A, Ozga JA, Lopes-Lutz D, Schieber A, Reinecke DM. Characterization of proanthocyanidins in pea (*Pisum sativum* L.), lentil (*Lens culinaris* L.), and faba bean (*Vicia faba* L.) seeds. Food Res Int 2012; 46: 528-35.
[http://dx.doi.org/10.1016/j.foodres.2011.11.018]

[151] Martinez-Micaelo N, González-Abuín N, Ardèvol A, Pinent M, Blay MT. Procyanidins and

inflammation: molecular targets and health implications. Biofactors 2012; 38(4): 257-65.
[http://dx.doi.org/10.1002/biof.1019] [PMID: 22505223]

[152]  González-Abuin N, Pinent M, Casanova-Marti A, Arola L, Blay M, Ardevol A. Procyanidins and their healthy protective effects against type 2 diabetes. Curr Med Chem 2015; 22(1): 39-50.
[http://dx.doi.org/10.2174/0929867321666140916115519] [PMID: 25245512]

[153]  López-Amorós ML, Hernández T, Estrela I. Effect of germination on legume phenolic compounds and their antioxidant activity. J Food Compos Anal 2006; 19: 277-83.
[http://dx.doi.org/10.1016/j.jfca.2004.06.012]

[154]  Amarowicz R, Estrella I, Hernández T, Robredo S, Troszyńska A. Kosińska a, Pegg RB. Free radical-scavenging capacity, antioxidant activity, and phenolic composition of green lentil (*Lens culinaris*). Food Chem 2010; 121: 705-11.
[http://dx.doi.org/10.1016/j.foodchem.2010.01.009]

[155]  Amarowicz R, Estrella I, Hernandez T, Troszyńska A. Antioxidant activity of extract of adzuki bean and its fractions. J Food Lipids 2008; 15: 119-36.
[http://dx.doi.org/10.1111/j.1745-4522.2007.00106.x]

[156]  Zhao Y, Du SK, Wang H, Cai M. *In vitro* antioxidant activity of extracts from common legumes. Food Chem 2014; 152: 462-6.
[http://dx.doi.org/10.1016/j.foodchem.2013.12.006] [PMID: 24444962]

[157]  Fernández-Marín B, Milla R, Martín-Robles N, *et al.* Side-effects of domestication: cultivated legume seeds contain similar tocopherols and fatty acids but less carotenoids than their wild counterparts. BMC Plant Biol 2014; 14: 1599.
[http://dx.doi.org/10.1186/s12870-014-0385-1] [PMID: 25526984]

[158]  Rao AV, Rao LG. Carotenoids and human health. Pharmacol Res 2007; 55(3): 207-16.
[http://dx.doi.org/10.1016/j.phrs.2007.01.012] [PMID: 17349800]

[159]  Ma L, Lin XM. Effects of lutein and zeaxanthin on aspects of eye health. J Sci Food Agric 2010; 90(1): 2-12.
[http://dx.doi.org/10.1002/jsfa.3785] [PMID: 20355006]

[160]  Fiedor J, Burda K. Potential role of carotenoids as antioxidants in human health and disease. Nutrients 2014; 6(2): 466-88.
[http://dx.doi.org/10.3390/nu6020466] [PMID: 24473231]

[161]  Priyadarshani AM. A review on factors influencing bioaccessibility and bioefficacy of carotenoids. Crit Rev Food Sci Nutr 2015. [Epub ahead of print]
[http://dx.doi.org/10.1080/10408398.2015.1023431] [PMID: 26168011]

[162]  Boschin G, Arnoldi A. Legumes are valuable sources of tocopherols. Food Chem 2011; 127(3): 1199-203.
[http://dx.doi.org/10.1016/j.foodchem.2011.01.124] [PMID: 25214114]

[163]  Reiter E, Jiang Q, Christen S. Anti-inflammatory properties of α- and γ-tocopherol. Mol Aspects Med 2007; 28(5-6): 668-91.
[http://dx.doi.org/10.1016/j.mam.2007.01.003] [PMID: 17316780]

[164]  Devaraj S, Leonard S, Traber MG, Jialal I. Gamma-tocopherol supplementation alone and in

combination with alpha-tocopherol alters biomarkers of oxidative stress and inflammation in subjects with metabolic syndrome. Free Radic Biol Med 2008; 44(6): 1203-8.
[http://dx.doi.org/10.1016/j.freeradbiomed.2007.12.018] [PMID: 18191645]

[165] Gohil K, Vasu VT, Cross CE. Dietary α-tocopherol and neuromuscular health: search for optimal dose and molecular mechanisms continues! Mol Nutr Food Res 2010; 54(5): 693-709.
[http://dx.doi.org/10.1002/mnfr.200900575] [PMID: 20187127]

[166] Vardi M, Levy NS, Levy AP. Vitamin E in the prevention of cardiovascular disease: the importance of proper patient selection. J Lipid Res 2013; 54(9): 2307-14.
[http://dx.doi.org/10.1194/jlr.R026641] [PMID: 23505320]

[167] Silva LR, Pereira MJ, Azevedo J, *et al.* Inoculation with Bradyrhizobium japonicum enhances the organic and fatty acids content of soybean (Glycine max (L.) Merrill) seeds. Food Chem 2013; 141(4): 3636-48.
[http://dx.doi.org/10.1016/j.foodchem.2013.06.045] [PMID: 23993531]

[168] Harris WS, Mozaffarian D, Rimm E, *et al.* Omega-6 fatty acids and risk for cardiovascular disease: a science advisory from the American Heart Association Nutrition Subcommittee of the Council on Nutrition, Physical Activity, and Metabolism; Council on Cardiovascular Nursing; and Council on Epidemiology and Prevention. Circulation 2009; 119(6): 902-7.
[http://dx.doi.org/10.1161/CIRCULATIONAHA.108.191627] [PMID: 19171857]

[169] Ramsden CE, Faurot KR, Carrera-Bastos P, Cordain L, De Lorgeril M, Sperling LS. Dietary fat quality and coronary heart disease prevention: a unified theory based on evolutionary, historical, global, and modern perspectives. Curr Treat Options Cardiovasc Med 2009; 11(4): 289-301.
[http://dx.doi.org/10.1007/s11936-009-0030-8] [PMID: 19627662]

[170] Mocellin MC, Camargo CQ, Nunes EA, Fiates GM, Trindade EB. A systematic review and meta-analysis of the n-3 polyunsaturated fatty acids effects on inflammatory markers in colorectal cancer. Clin Nutr 2015; pii: S0261-5614(15): 00123-5.
[http://dx.doi.org/10.1016/j.clnu.2015.04.013] [PMID: 25982417]

# CHAPTER 2

# Bioactive Compounds from *Brassicaceae* as Health Promoters

**Nieves Baenas[1,†], Marta Francisco[2,†], Pablo Velasco[2], María Elena Cartea[2], Cristina García-Viguera[1], Diego A. Moreno[1,*]**

[1] *Phytochemistry Lab, Food Sci. & Technology Dept., CEBAS-CSIC, Murcia, Spain*

[2] *Group of Genetics, Breeding and Biochemistry of Brassicas, MBG-CSIC, Pontevedra, Spain*

[†] *These authors contributed equally to this work*

**Abstract:** This work provides an up to date review of the information available about bioactive compounds present in the *Brassicaceae* family (glucosinolates, phenolics and vitamins) in relation to human health. The *Brassicaceae* plant family includes a large variety of species and cultivars, some of the most known are *Brassica oleracea* (*e.g.* broccoli, cabbage, Brussels sprouts), *Brassica rapa* (*e.g.* turnips), *Brassica napus* (*e.g.* rapeseed), Raphanus sativus (radishes), and Sinapis alba (mustards). In the recent years, these crops are increasingly consumed for possible health benefits as a good source of bioactive compounds. The sulphur containing compounds glucosinolates are almost exclusively found in this family, being their beneficial health effect supposed to be induced by their hydrolysis products, the isothiocyanates. In *in vitro* (human cell lines) and *in vivo* studies (animal models and human intervention assays) isothiocyanates have demonstrated their protective effects in carcinogenesis, chronic inflammation and neurodegeneration. The phenolic compounds mainly studied are flavonols, anthocyanins and hydroxycinnamic acids, which principal bioactivity is their antioxidant capacity. The carotenoids β-carotene, lutein and zeaxanthin, as well as, vitamins C, E and K have also been considered as nutrients with biological activity. The phytochemical wealth of *Brassica* foods is gathering attention from the scientific community for being potentially protective for the cardiovascular system and against certain types of cancer, and neurological disorders, mainly because of their anti-inflammatory and antioxidant properties.

**\* Corresponding author Diego A. Moreno:** C.E.B.A.S.-C.S.I.C. – FOOD SCI. & TECHNOL. DEPT., Campus de Espinardo – Edificio 25, P.O. BOX 164, E-30100 Espinardo, Murcia (SPAIN); Tel: +34 968 39 6369; Fax: +34 968 39 6213; Email: dmoreno@cebas.csic.es.

**Luís Rodrigues da Silva and Branca Maria Silva (Eds.)**

Even it is not yet possible to recommend a particular "daily dose" for human consumption of cruciferous foods for disease prevention, there is growing evidence regarding the protective effects of *Brassica* bioactive compounds for health via regulation of signaling pathways and cellular metabolism

**Keywords:** Antiinflammatory, Antioxidant, *Brassicaceae*, Cardiovascular disease, Carotenoids, Chemoprevention, Cruciferous, Glucosinolates, Isothiocyanates, Minerals, Neurodegeneration, Phenolic compounds, Vitamins.

## INTRODUCTION

*Brassicaceae* family, commonly termed the mustard family or Cruciferae, represents a monophyletic group including approximately 350 genera and 3,700 species, which has been the subject of much scientific interest, with many crops of socioeconomical relevance (food and spices, condiments, oils), forage or ornamental. This family includes common species of food staples such as: broccoli, cauliflower, Brussels sprouts, cabbages, belonging to *Brassica* oleracea; turnips and Chinese cabbages of *Brassica* rapa; oilseeds of *Brassica* napus (rapeseed, leaf rape); mustards (*Sinapis alba*); and radishes (*Raphanus sativus*), among others. *Brassicaceae* crops are dated in Europe and northern Asia for at least 600 years and in the earlier part of the 20th century they have grown in North America, with productions in Europe around the 70 million tons/annum [1].

*Brassicaceae* crops are widely distributed in the World: Southwestern and Central Asia, Mediterranean Europe, and North and South America. *Brassica* production and consumption has increased worldwide in the last years, but only from a few cultivated genera [2]. There are numerous further species with great potential for exploitation in 21st century agricultural and food commodities, particularly as sources of bioactive phytonutrients.

## PHYTONUTRIENTS IN CRUCIFEROUS PLANTS AND FOODS

The *Brassicaceae* vegetables have been widely studied for their beneficial effects on human health through epidemiological studies [3], being nutritive foods rich in essential nutrients and phytochemicals that may act synergistically in the food matrix to modulate the cell metabolism and help in the prevention and treatment of certain types of cancer, cardiovascular health problems, and neurodegenerative conditions of the aging human being (Table **1**) [4]. Although vegetable cruciferous plants are sources of fiber, folate, vitamins (A, E, C, and K) and minerals (Ca, Fe, K, Cu, Zn, P, Mn, and Mg, among others), the major body of evidence in the scientific literature is concentrated in the contents of secondary metabolites, such as flavonoids and carotenoids, and specially glucosinolates (GLSs). These compounds are mainly present in this family and are hydrolyzed to isothiocyanates (ITCs), which may be responsible of the chemoprotective activity and the reduction in the risk of suffering a number of cancers associated with the intake of cruciferous foods. Also the health-promoting effects of crucifers have been attributed at least in part to their bioactive composition rich in natural antioxidants, such as vitamins (C, A, E, K, *etc.*), carotenoids and phenolic compounds [5].

**Table 1. Nutrients and phytochemicals presents in cruciferous plants and their physiological functions.**

| Compounds and chemical structures | Physiological functions | References |
|---|---|---|
| GLSs and ITCs Glucosinolate        myrosinase        Isothiocyanate | Induction of detoxification enzymes Apoptosis and arrest of tumor cell growth Decrease adipogenesis and inflammation Reduce oxidative stresses | [4, 6 - 8] |
| Flavonols  Quercetin — OH / Kaempferol — H / Isorhamnetin — OCH$_3$ | Prevent the oxidation of LDL Capillary protective effect Reduce serum levels of glucose Tumor inhibitory effect Anti-inflammatory, antimicrobial and anti-allergic | [5, 9, 10] |

*(Table 1) contd.....*

| Compounds and chemical structures | Physiological functions | References |
|---|---|---|
| Anthocyanins<br><br><br><br>Cyanidin-3-(sinapoyl)diglucoside-5-glucoside | Antioxidant power and antigenotoxic | [11] |
| Hydroxycinnamic acids<br><br> | Cellular defense of peroxynitrite-mediated disorders | [12, 13] |
| Vitamins and β-carotene<br><br><br>Vitamin C            β-carotene<br><br><br>Vitamin E | Protection against free radicals<br>Cytoprotective functions<br>Preserve protein integrity | [14 - 17] |
| Minerals Fundamentally the elements or Mainly the elements K, Ca, Na, Mg, Fe, Zn, Se and Mn. | Participation in metabolic activities Biochemical and nutritional functions | [3, 18] |

Hydroxycinnamic acids table:

|  | $R_1$ | $R_2$ |
|---|---|---|
| *P*-Coumaric | H | H |
| Caffeic | OH | H |
| Ferulic | $OCH_3$ | H |
| Sinapic | $OCH_3$ | $OCH_3$ |

## Glucosinolates and Bioactive Isothiocyanates

The GLSs are secondary metabolites, sulphur and nitrogen-containing compounds with a common structure which comprises a $\beta$-D-thioglucose group, a sulphonated oxime moiety, and a variable aglycone side-chain derived from natural amino acids, that determine the final structure, being mainly derived from methionine, tryptophan or phenylalanine. Therefore, GLSs can be classified by their precursor amino acids as aliphatic (derived from alanine, leucine, methionine or valine), aromatic (from phenylalanine or tyrosine) and indolic (from tryptophan). The studies of the profile of GLSs indicate significant differences among species, according to the type and intensity of environmental stress, growth conditions and storage, processing and cooking methods [3, 19 - 21]. The GLSs load in plant tissues is highly variable, being seeds the plant part with the highest contents, followed by the germinating seeds and sprouts –that may present a 10-fold increase compared to commercial inflorescences or heads from adult plants– and generally followed by leaves and roots. This amount of GLSs may range from 1% to 10% (on a dry weight basis) in the seeds of some species [2].

GLSs are hydrolyzed to ITCs, their biologically active hydrolysis metabolites, both by the action of the enzyme myrosinase (thioglucoside glycohydrolase EC EC:3.2.1.147), which comes into contact with GLSs when there is a tissue disrupted by crushing or herbivory/chewing or by the action of the gut microflora upon human ingestion. Intact GLSs have no known biological activity; thus, the bioavailability of bioactive hydrolysis products is dependent not only on ingestion of GLSs, but also on their conversion prior to passage across the gut wall [6]. Inactivation of the plant myrosinase also decreases bioavailability of ITCs because the enzyme is heat sensitive, as occurs when *Brassica* vegetables are ingested cooked, thus boiling or steaming for more than 3-5 min and blanching previous frozen production will lead to loss of its activity [22].

The bioavailability of GLSs is measured by analyzing the mercapturic acid pathway products (mercapturates) which acts as a bioindicator or marker of intake and also are useful to study the bioaccesibility and bioavailability of the GLSs breakdown products, the ITCs, which gives rise to N-acetyl-cysteine conjugates. *In vitro* and *in vivo* studies mainly focused in the use of sulforaphane from broccoli (SFN), has shown the influence of this bioactive compound on the cellular cytoprotective mechanisms involved in all the stages of development of cancer, through the selective induction of detoxification Phase II enzymes, to detoxify the products (electrophilic metabolites) of the activity of phase I enzymes to avoid the damage on the DNA (glutathione S-transferases, UDP-glucuronosyl transferases, and quinone reductase) [23] and through the limitation of the activity of Phase I enzymes [24]. A diet of 3-5 servings per week is sufficient to cause a 30% or 40% decrease in risk for a number of cancers [25].

Glucoraphanin (GRA) is the GLS precursor of the bioactive ITC SFN, which is being widely studied as a potent protector of carcinogenesis. Histone deacetylases (HDAC), which remove acetyl groups from proteins, has been studied recently and SFN metabolites were reported as inhibitors altering their gene expression and protein function [26]. One major step to determine the absorption of SFN is the hydrolysis of GRA by the action of the enzyme myrosinase. When comparing the intake of supplements rich in GRA with inactivated myrosinase, against the intake of fresh broccoli sprouts with the active enzyme, Clarke *et al.*, observed a much limited SFN absorption in healthy adults (7-fold lower) [27]. Also Vermeulen *et al.*, observed higher excretion of SFN metabolites after consumption of raw *versus* cooked broccoli – by 11% [28]. Other interesting work showed in plasma and urine higher levels of total SFN metabolites (3-5 times) in fresh broccoli sprouts consumers compared to myrosinase-treated broccoli sprouts extract containing SFN but not GRA; therefore, GRA hydrolysis to produce SFN is not the only one factor affecting the absorption of SFN, other compounds present in the broccoli sprout food matrix, such as minerals, vitamins, other nutrients and phytochemicals and fiber may facilitate the transport of SFN across cell membranes [29]. In SFN bioavailability, also the total amount of SFN estimated could derive from the interconversion of erucin, from the GLS glucoerucin, to SFN *in vivo* [8].

Not only the ITC SFN, but also erucin, from the precursor GLS glucoerucin, iberin from glucoiberin, sulforaphene from glucoraphenin (which differs from GRA by a double bond), phenethyl ITC from gluconasturtiin, and indole--carbinol (I3C) from indole GLSs (4-Hydroxi-, 4-Methoxy-, Neo- and glucobrassicin), have been studied because their bioactivity, triggering the transcription factor Nrf2 into de nucleus, where the antioxidant response element (ARE) promoter region activate multiple genes, including both phase II detoxification enzymes and several antioxidant enzymes, among others, and induce cell cycle arrest and apoptosis [30].

Recent studies have shown that the I3C plays important roles in apoptosis and arrest of cell growth in breast and prostate cancer cells [31], and may potential benefits in preventing obesity and its comorbidities through different mechanisms including the reduction of adipogenesis and inflammation, and the increased thermogenesis [32].

Further *in vitro* and *in vivo* assays to understand GLSs bioavailability, would encourage the use of cruciferous vegetables as preventive and health food within the confines of animal studies or human trial for any form of cancer.

## Phenolic Compounds

Phenolic compounds are ubiquitous phytochemicals in plants and plant foods (more than 8,000 described, characterized by having at least one aromatic ring with one or more hydroxyl groups attached). The structure of phenolic compounds may be very simple and with low molecular-weight, with single aromatic-ringed compounds or very complex (*i.e.*, tannins and other (poly)phenolics) [33, 34]. These compounds perform a variety of functions in the plant, generally centered on responses to pathogen attacks, UV protection, colour and sensory characteristics [35]. The phenolic compounds may be classified according to their number and arrangements of their carbon atoms in flavonoids (flavonols, flavones, flavan-3-ols, anthocyanidins, flavanones, isoflavones and others) and non-flavonoids (phenolic acids, hydroxycinnamates, stilbenes and others) [36].

Diets rich in plant-derived foods rich in phenolic compounds, such as those from the cruciferous family, have been reported to exert health-promoting benefits at different levels: anti-inflammatory, enzyme inhibition, antimicrobial, antiallergic, vascular and cytotoxic antitumor activity, and the most widely cited action of phenolics, their antioxidant activity [37, 38]. Phenolic compounds can play important roles in scavenging free radicals and up-regulation of certain metal-chelation reactions. The reactive oxygen species (ROS, singlet oxygen, peroxynitrite, hydrogen peroxide), must be eliminated from cells to maintain healthy metabolic functions, and such reductions are positively associated with the ion transport systems and may affect the redox signaling in the cells [34]. Despite the beneficial effects of phenolic compounds it must be taken into account that only a small percentage of the dietary phenolics get inside the cells and are absorbed and metabolized. The plasma concentrations after intake of polyphenol-rich foods depend on the food source, and the polyphenol chemistry, and for example, it may vary from 0.3–0.75 µmol/L after consumption of 80–100 mg quercetin equivalents [39]. Moreover, (poly)phenolics are modified during their metabolism in the gastrointestinal tract and these modifications involve conjugation to produce glucuronides or sulphate conjugates by intestinal and/or hepatic detoxification enzymes. However, the major part of these molecules is metabolized by the colonic microflora rendering the so called microbial metabolites. Those microbial metabolites can be analysed in blood (plasma) and urine extractions, after ingestion, but only very small fraction of non-conjugated phenols in their original form can be found. This implies that these microbial metabolites rather than the native phenolics are responsible for the beneficial biological effects in the body [33].

*Brassica* vegetables are generally rich in polyphenols, although the profile and content of those compounds in the plant may vary depending *e.g.* on climatic conditions and harvest season [14, 40 - 42]. Moreover, differences in phenolic content can be expected between different cultivars as well as within plant organs [43]. Food processing may also affect phenolic content [44 - 47]. Phenolic contents may vary from 15.3 mg·100g$^{-1}$ in white cabbage (on a fresh weight basis), to 337 mg·100g$^{-1}$ in broccoli. Cartea and co-workers [48] extensively reviewed the phenolic profiles in many different species of *Brassicaceae* and the most widespread and diverse group of polyphenols in these species are the flavonoids (mainly flavonols, but also anthocyanins) and the phenolic acids.

### Flavonoids

Flavonoids are low molecular weight plant-secondary metabolites, consisting of 15 carbon atoms with two aromatic rings (A and B), connected by a three-carbon bridge (C6–C3–C6) configuration that usually in the form of a C-heterocyclic ring. The flavonoids described in *Brassica* spp. are *O*-glycosides of quercetin, kaempferol and isorhamnetin. Besides, they may be conjugated with different organic acids, most frequently at the 3-position of the C-ring, but substitutions may be also placed in the 5, 7, 4′, 3′ and 5′ positions [33]. To date, more than 20 flavonols have been described in *Brassica* vegetables such as kale, white cabbage, cauliflower, and broccoli as well as in *B. napus* and *B. rapa* leaves. Among them, the main flavonols were identified as kaempferol and quercetin 3-*O*-sophoroside-7-*O*-glucosides and combinations with hydroxycinnamic acids (*i.e.*, kaempferol and quercetin 3-*O*-(caffeoyl/sinapoyl)-sophoroside-7-*O*-glucoside). In the *B. rapa* group, in addition to quercetin and kaempferol derivates, it can be found derivatives of the flavonol isorhamnetin. In cruciferous varieties for fresh-cut and baby leaf supply including *Diplotaxis erucoides* L., *D. tenuifolia* L., *Eruca sativa* L., *Bunias orientalis* L., and *Nasturtium officinale*, quercetin and kaempferol glycosylated derivatives are the major flavonols [38, 48].

The glycosylated flavonols (3-sophoroside-7-glucosides of kaempferol) are receiving more attention in terms of beneficial health effects such as reduction in the risk of suffering certain age-related chronic health problems [38]. Accordingly, quercetin glycosides found at high concentrations in broccoli displayed the ability to prevent the oxidation of LDL by scavenging free radicals and chelating transition metal ions [9]. As a result, quercetin glycosides may help against certain conditions relevant to adult health: cancer, atherosclerosis and chronic inflammation [9]. On the other hand, isorhamnetin glycosides in mustard leaves were named responsible for the hypoglycemic effect using an antioxidant capacity test [49].

Anthocyanins are also present in *Brassica* vegetables conferring red pigmentation in red cabbages, red radishes, purple cauliflowers and broccolis. The major anthocyanins identified in these crops are cyanidin derivatives. In red cabbage and broccoli sprouts the major anthocyanins identified were the cyanidin 3-*O*-(sinapoyl)(feruloyl) diglucoside-5-*O*-glucoside and the cyanidin 3-*O*-(sinapoyl)(sinapoyl) diglucoside-5-*O*-glucoside [50]. The biological activity of the anthocyanins have been related also with the antioxidant capacity and the antigenotoxic properties, also repairing the cytological injuries caused by $Cu^{2+}$ stress on lymphocytes [11].

## Phenolic Acids and Derivatives

The compounds present in the phenolic acids class are consisting in two groups, the hydroxybenzoic and the hydroxycinnamic acids. The hydroxybenzoates include gallic, *p*-hydroxybenzoic, protocatechuic, vanillic and syringic acids, which have the C6–C1 structure in common. In the hydroxycinnamic acids, the compounds are aromatic with a three-carbon side chain (C6–C3), being caffeic, ferulic, *p*-coumaric and sinapic acids the most common in *Brassica*s. It is common to find the phenolic acids conjugated also with sugars or with other hydroxycinnamic acids [48]. This phenolic class is abundant in *B. oleracea* crops, such as kale, cabbage, broccoli, and cauliflower. In these crops, hydroxycinnamoyl gentiobiosides (1-*O*-caffeoylgentiobiose and 1,2,6-tri-*O*-sinapoylgentiobiose) and hydroxycinnamoylquinic (5-caffeoyl quinic acid) acids are the major representatives [51].

The antioxidant scavenger properties of *Brassica*extracts rich in phenolic acids have also been proved *in vivo*. As a result, the intervention in human subjects with a diet rich in Brussels sprouts showed a reduction on DNA damage in terms of a decreased excretion 8-oxo-7,8-dihydro-2'-deoxyguanosine (8-oxodG) in urine [52]. It has been also reported that even with a short intervention, with broccoli sprouts in rats, a strong protection in the heart against oxidative stress and cell death caused by ischemia-reperfusion or diabetes, can be measured [12]. The sinapic acids, also present in high amounts in cruciferous foods, may also contribute to the cellular defense mechanisms avoiding peroxynitrite-mediated disorders [13].

## NUTRIENTS: MINERALS AND VITAMINS

## Minerals

The essential minerals (Na, K, Ca, Mg, Cl and P) are required in high amounts in diet (>50 mg/day), while the metals and trace elements (Fe, Zn, Cu, Mn, I, F, Se, Cr, Mo, Co, Ni) are needed in much lower concentrations (<<50 mg/day). The mineral nutrients are involved in different life processes, *e.g.* as electrolytes, as enzyme constituents, building materials in bones and teeth, *etc.* [18]. The microelements content in cruciferous foods are fundamentally K, Ca, Na, Mg, Fe,

Zn, Se and Mn. Ready-to-eat cruciferous sprouts are an excellent source of these compounds, showing higher concentrations of minerals than seeds (12-45% higher according to the compound) [53]. The content of the main minerals could vary among species and crops. Broccoli, cauliflower, turnip, cabbage, red cabbage and rutabaga show values ranging 170-300 mg·100g$^{-1}$ (on a fresh weight basis of K in raw products, being the predominant mineral in crucifers. The contents of P, Ca, Na and Mg range 10 - 60 mg·100g$^{-1}$ (on a fresh weight basis), and for Fe and Zn, 0.3 - 0.8 mg·100g$^{-1}$ F.W. [54]. Seasonal variations could affect the content of minerals, such as Fe, Ca and Zn, being higher in wet season than dry season [55]. Broccoli can be suggested as a 'good source' of Ca and Mg for human nutrition, with comparable bioavailability to that of milk, and therefore, may be considered an important alternative source of Ca in those population groups with limited access or intake of dairy products [56]. Different cooking methods (boiling, steaming, microwaving and frying) not affected significantly to the mineral content of broccoli florets; therefore, on average, an edible portion (200g of raw broccoli) could provide, over 20% of the daily requirements of minerals [21, 57].

**Vitamins and Carotenoids**

The vitamins present in cruciferous vegetables are: vitamin C, A, E, B and K, thiamin, riboflavin, niacin and folate. The major natural antioxidants in cruciferous foods are the vitamins (C, E, *etc.*), the carotenoids, and the (poly)phenolics [5]. The variation in the contents of these intrinsic antioxidants is caused by many factors: variety, organ and maturity at harvest, soil conditions and agricultural practices, post-harvest management, industrial and domestic processing, inducing all, many differences in the health-promoting properties of these vegetables when reaching the plate [58, 59]. These nutrients and phytochemicals are radical scavengers that inhibit the chain initiation or break the chain propagation (the second defense line) of the oxidative stress reactions. Diverse studies have shown a synergetic effect of both hydrosoluble (vitamin C) and lipid-soluble antioxidants (carotenoids and vitamin E), as in combinations of α-tocopherol or vitamin C plus phenolic compounds [60].

Vitamin C (Ascorbic Acid and Dehydroascorbic Acid), has many biological roles in human physiology, through its protective effects against free radicals, prevention of DNA mutation in the cells, as well as the protection against lipid peroxidative damage, and repairing amino acid residues to preserve the protein integrity [58, 61]. These mechanisms of actions involve the prevention of certain cancer and cardiovascular diseases [62, 63]. For instance, vitamin C was established as responsible of 10–12% of the total antioxidant capacity of broccoli or cabbage [5]. The vitamin C can not be synthesized in the human body and therefore needs to be taken through diet. *Brassica* vegetables generally contain high amounts of vitamin C, and depending on dietary habits and geographical locations, may provide up to 50% of the daily RDI (recommended dietary intake) for human adults [58]. Among species, the vegetables from the genera *Brassica* (such as broccoli, red cabbage, Brussels sprouts, and kale) exhibit higher content of this vitamin (ranging 50-200 mg·100g$^{-1}$ F.W) than other species, depending distinct plant organs and physiological stages [54]. Also cooking methods decrease the Vitamin C content, causing steaming the lowest loss comparing to microwave and boiling [58].

On the vitamin E group, $\alpha$-tocopherol is the most common and biologically active form, and like all essential nutrients, a minimum level of vitamin E is also essential for wellness and health. It reduces the peroxyl radicals produced from polyunsaturated fatty acids (PUFAs) in phospholipidic membranes or lipoproteins. The severe deficiency of vitamin E results in various neurological problems including ataxia (impaired balance and coordination), myopathy (muscle weakness) and damage to the retina. Suboptimal dietary intakes (or plasma levels of vitamin E below normal) are associated with increased risk of cardiovascular disease, some cancers and decreased immune function [15, 64].

Vitamin K, being phylloquinone the major dietary form, is found ranging 15-100 µg·100g$^{-1}$ F.W in common *Brassica* vegetables, such as broccoli, cabbage, red cabbage and cauliflower, and act as cofactor for the enzyme $\gamma$-glutamyl carboxylaseis, involved in the blood coagulation cascade and catalysis of the carboxylation of osteocalcin in bone [65].

Thiamine (B1) and riboflavin (B2) have been studied in cruciferous sprouts, radish, rapeseed and white mustard seeds contain vitamin B1 (0.41-0.70 mg·100g$^{-1}$ D.W.); however, its amount found in the ready-to-eat sprouts were 40% lower.

In contrast, the content of vitamin B2 in ready-to-eat sprouts were 3 -fold higher when compared to the seeds (0.096-0.138 mg·100g$^{-1}$ D.W.).

*Brassica* crops show high levels of folate (15-60 µg·100g$^{-1}$ F.W.), which is a scarce and important vitamin related to the reduced risk of vascular diseases, cancer and neural tube defects [3].

Carotenoids (carotenes and xanthophylls) are yellow, orange, and red lipid-soluble compounds characteristic of many fruits and vegetables. Leafy *Brassica*s are sources of carotenoids, particularly lutein, zeaxanthin and *β*-carotene [54]. These compounds are also reported for antioxidant functions such as quenching of singlet oxygen and other electronically charged molecules produced in reactions after photo or chemical excitation and they also react with peroxyl or alkoxyl radicals. The carotenoids are precursors of vitamin A (*i.e.β*-carotene, *γ*-carotene, and *β*-cryptoxanthin), and due to their conjugated double bonds they are both radical scavengers and quenchers of singlet oxygen [66, 67]. Brussels sprouts (6.1 mg·100g$^{-1}$ F.W.), broccoli (1.6 mg·100g$^{-1}$ F.W.), red cabbage (0.43 mg·100g$^{-1}$ F.W.), and white cabbage (0.26 mg·100g$^{-1}$ F.W.) are the species with higher content of these compounds. Several studies have demonstrated that carotenoids significantly down-regulated the expression of pro-inflammatory cytokines, possibly due to alterations of the NF-κB pathway and impacted Nrf2, a transcription factor related to the expression of detoxifying enzymes, in addition to directly quenching ROS, all related to the cytoprotective systems that reduce the risk for cardiovascular diseases and different types of cancer [68].

Even *Brassicaceae* vegetables are a good source of antioxidants, the potential health benefits mainly depend on the genotype and subspecies, as after studying different varieties and cultivars of cabbage, Chinese cabbage, cauliflower, broccoli and Brussels sprouts, the major source of Vitamin C (52.9 mg·100g$^{-1}$ F.W.), β-carotene (0.81 mg·100g$^{-1}$ F.W.), lutein (0.68 mg·100g$^{-1}$ F.W.), DL-*α*-tocopherol (vitamin E) (0.47 mg·100g$^{-1}$ F.W.) and phenolics (63.4 mg·100g$^{-1}$ F.W.) was represented by broccoli [69].

## OTHER NUTRIENTS

The vegetables of the cruciferous family are good sources of other additional macronutrients. When comparing *Brassica* vegetables with other plant foods of high water content, the levels of fiber are relatively high. In white cabbage more than 30% of its total carbohydrate content is made from dietary fiber. Similarly, and depending on the crop, they can also represent significant sources of amino acids and proteins. Two hundred calories of steamed broccoli will provide 20g of protein [70]. Raw broccoli, cabbages and cauliflowers also contain folates, a relatively scarce and relevant nutrient that acts as a coenzyme in the synthesis of DNA, RNA and protein components, as well as in many single carbon transfer reactions [3]. Recently, it has been described that germinating seeds or prouts from cruciferous varieties can be also a good source of other antioxidants such us melatonin and serotonin [71]. As a conclusion, regular dietary *Brassica*vegetables it may account for an important promising chemopreventive dietary constituents (GLSs, vitamins, phenolics, minerals, fiber, *etc.*) which may protect the cell systems against free radical damages, LDL oxidation, pathogenesis of cardiovascular problems, and the DNA damages leading to cancer processes.

## FUTURE PERSPECTIVES

The increasing awareness among scientists, food manufacturers, more and more health-conscious consumers worldwide, on the beneficial effects of the *Brassicaceae* phytochemical-rich foods, has prompt the production of these vegetables in sustainable practices. From the plant genetics approaches, through traditional breeding programs or through bioengineering of the secondary metabolism, to induce the accumulation of a particular nutrient or phytochemical, not many advatages have been reported. To the best of our knowledge, breeding programs to increase or decrease the content of a particular phenolic compound related to human health with horticultural *Brassica* crops have not been carried out. However, the modification of the synthesis of GLSs is being currently carried out in different crops such as broccoli. Through the introgression of a *Brassica villosa* MYB28 allele, that enhances sulphate assimilation and accumulates methionine-derived GLSs, it was developed commercially broccoli F1 hybrids with increased concentrations of GRA [72]. These investigations on increasing levels of *Brassica* phytochemicals may have a potential for human intervention studies to investigate the effects of a specific compound on human health.

On the other hand, as a result of the increasing available information about the health-promoting properties of cruciferous plants, there is an increasing demand of foods and food products enriched in *Brassica* bioactives. Minimally processed broccoli byproducts can be used as a source of bioactive ingredients, mainly GSLs and phenolic compounds to design novel beverages. A squeezed liquid composition is described by Kumazawa [73] that is rich in GLSs and has a good balance of vitamin C, colour, flavour and similar attributes. Moreover, the use of plant ingredients rich in *Cruciferae* bioactive compounds as functional foods and ingredients provide additional routes for the industrial exploitation of these attractive natural plant products. Only recently, pharmaceutical forms (pills, powders, capsules, vials, *etc.*) containing GLSs as active principles (commonly, broccoli extracts claiming sulforaphane presence in the formulation) have appeared in the markets, even with mixes of varieties of sources to supply SFN and I3C. In a very close future, a better understanding of the bioavailability, metabolism and physiological relevance of these dietary bioactive compounds and the intervention of the gut microbiota in the relationship will all help to elucidate the mechanisms by which these compounds are useful and suitable tools in the dietary interventions to treat and manage diseases as well as the maintenance of a wellbeing status in animals and the human beign.

## CONFLICT OF INTEREST

The authors confirm that they have no conflict of interest to declare for this publication.

## ACKNOWLEDGEMENTS

This work was supported by the Spanish Ministry of Economy and Competitiveness through the projects: AGL2012-35539 and AGL2013-46247-P.

## REFERENCES

[1]     Schmidt R, Bancroft J. Genetics and genomics of the Brassicaceae. Germany: Springer 2011.
[http://dx.doi.org/10.1007/978-1-4419-7118-0]

[2]     Fahey JW, Zalcmann AT, Talalay P. The chemical diversity and distribution of glucosinolates and isothiocyanates among plants. Phytochemistry 2001; 56(1): 5-51.
[http://dx.doi.org/10.1016/S0031-9422(00)00316-2] [PMID: 11198818]

[3]     Jahangir M, Kim HK, Choi YH, Verpoorte R. Health-affecting compounds in *Brassicaceae.* Com Rev Food Sci F 2009; 8: 31-43.
[http://dx.doi.org/10.1111/j.1541-4337.2008.00065.x]

[4]     Dinkova-Kostova AT, Kostov RV. Glucosinolates and isothiocyanates in health and disease. Trends Mol Med 2012; 18(6): 337-47.
[http://dx.doi.org/10.1016/j.molmed.2012.04.003] [PMID: 22578879]

[5]     Podsędek A. Natural antioxidants and antioxidant capacity of Brassica vegetables: A review. Food Sci Technol (Campinas) 2007; 40: 1-11.

[6]     Angelino D, Jeffery E. Glucosinolate hydrolysis and bioavailability of resulting isothiocyanates: Focus on glucoraphanin. J Funct Foods 2014; 7: 67-76.
[http://dx.doi.org/10.1016/j.jff.2013.09.029]

[7]     Cartea ME, Velasco P. Glucosinolates in Brassica food: bioavailability in food and significance for human health. Phytochem Rev 2008; 7: 213-29.
[http://dx.doi.org/10.1007/s11101-007-9072-2]

[8]     Clarke JD, Hsu A, Riedl K, *et al.* Bioavailability and inter-conversion of sulforaphane and erucin in human subjects consuming broccoli sprouts or broccoli supplement in a cross-over study design. Pharmacol Res 2011; 64(5): 456-63.
[http://dx.doi.org/10.1016/j.phrs.2011.07.005] [PMID: 21816223]

[9]     Ackland ML, van de Waarsenburg S, Jones R. Synergistic antiproliferative action of the flavonols quercetin and kaempferol in cultured human cancer cell lines. In Vivo 2005; 19(1): 69-76.
[PMID: 15796157]

[10]    Yokozawa T, Kim HY, Cho EJ, Choi JS, Chung HY. Antioxidant effects of isorhamnetin 3,7-di---β-D-glucopyranoside isolated from mustard leaf (*Brassica juncea*) in rats with streptozotocin-induced diabetes. J Agric Food Chem 2002; 50(19): 5490-5.
[http://dx.doi.org/10.1021/jf0202133] [PMID: 12207497]

[11]    Posmyk MM, Janas KM, Kontek R. Red cabbage anthocyanin extract alleviates copper-induced cytological disturbances in plant meristematic tissue and human lymphocytes. Biometals 2009; 22(3): 479-90.
[http://dx.doi.org/10.1007/s10534-009-9205-8] [PMID: 19152114]

[12]    Kataya HA, Hamza AA. Red cabbage (*Brassica oleracea*) ameliorates diabetic nephropathy in rats. Evid Based Complement Alternat Med 2008; 5(3): 281-7.
[http://dx.doi.org/10.1093/ecam/nem029] [PMID: 18830445]

[13]    Zou Y, Kim AR, Kim JE, Choi JS, Chung HY. Peroxynitrite scavenging activity of sinapic acid (3,5-dimethoxy-4-hydroxycinnamic acid) isolated from *Brassica juncea.* J Agric Food Chem 2002; 50(21): 5884-90.
[http://dx.doi.org/10.1021/jf020496z] [PMID: 12358454]

[14]    Borowski J, Szajdek A, Borowska EJ, Ciska E, Zielinski H. Content of selected bioactive components and antioxidant properties of broccoli (*Brassica oleracea* L.). Eur Food Res Technol 2008; 226: 459-65.
[http://dx.doi.org/10.1007/s00217-006-0557-9]

[15]   Traber MG, Frei B, Beckman JS. Vitamin E revisited: do new data validate benefits for chronic disease prevention? Curr Opin Lipidol 2008; 19(1): 30-8.
[PMID: 18196984]

[16]   Girard-Lalancette K, Pichette A, Legault J. Sensitive cell-based assay using DCFH oxidation for the determination of pro- and antioxidant properties of compounds and mixtures: Analysis of fruit and vegetable juices. Food Chem 2009; 115: 720-6.
[http://dx.doi.org/10.1016/j.foodchem.2008.12.002]

[17]   Nagao A. Absorption and function of dietary carotenoids. Forum Nutr 2009; 61: 55-63.
[http://dx.doi.org/10.1159/000212738] [PMID: 19367110]

[18]   Moreno DA, Carvajal M, López-Berenguer C, García-Viguera C. Chemical and biological characterisation of nutraceutical compounds of broccoli. J Pharm Biomed Anal 2006; 41(5): 1508-22.
[http://dx.doi.org/10.1016/j.jpba.2006.04.003] [PMID: 16713696]

[19]   Baenas N, Moreno DA, García-Viguera C. Selecting sprouts of *brassicacea*e for optimum phytochemical composition. J Agric Food Chem 2012; 60(45): 11409-20.
[http://dx.doi.org/10.1021/jf302863c] [PMID: 23061899]

[20]   Björkman M, Klingen I, Birch AN, *et al.* Phytochemicals of *Brassicaceae* in plant protection and human health--influences of climate, environment and agronomic practice. Phytochemistry 2011; 72(7): 538-56.
[http://dx.doi.org/10.1016/j.phytochem.2011.01.014] [PMID: 21315385]

[21]   Moreno DA, López-Berenguer C, García-Viguera C. Effects of stir-fry cooking with different edible oils on the phytochemical composition of broccoli. J Food Sci 2007; 72(1): S064-8.
[http://dx.doi.org/10.1111/j.1750-3841.2006.00213.x] [PMID: 17995900]

[22]   Tiwari U, Sheehy E, Rai D, Gaffney M, Evans P, Cummins E. Quantitative human exposure model to assess the level of glucosinolates upon thermal processing of cruciferous vegetables. LWT -. Food Sci Technol (Campinas) 2015; 63: 253-61.

[23]   Munday R, Munday CM. Induction of phase II detoxification enzymes in rats by plant-derived isothiocyanates: comparison of allyl isothiocyanate with sulforaphane and related compounds. J Agric Food Chem 2004; 52(7): 1867-71.
[http://dx.doi.org/10.1021/jf030549s] [PMID: 15053522]

[24]   Clarke JD, Dashwood RH, Ho E. Multi-targeted prevention of cancer by sulforaphane. Cancer Lett 2008; 269(2): 291-304.
[http://dx.doi.org/10.1016/j.canlet.2008.04.018] [PMID: 18504070]

[25]   Jeffery EH, Keck AS. Translating knowledge generated by epidemiological and *in vitro* studies into dietary cancer prevention. Mol Nutr Food Res 2008; 52 (Suppl. 1): S7-S17.
[PMID: 18327874]

[26]   Myzak MC, Tong P, Dashwood W-M, Dashwood RH, Ho E. Sulforaphane retards the growth of human PC-3 xenografts and inhibits HDAC activity in human subjects. Exp Biol Med (Maywood) 2007; 232(2): 227-34.
[PMID: 17259330]

[27]   Clarke JD, Riedl K, Bella D, Schwartz SJ, Stevens JF, Ho E. Comparison of isothiocyanate metabolite

levels and histone deacetylase activity in human subjects consuming broccoli sprouts or broccoli supplement. J Agric Food Chem 2011; 59(20): 10955-63.
[http://dx.doi.org/10.1021/jf202887c] [PMID: 21928849]

[28]  Vermeulen M, van den Berg R, Freidig AP, van Bladeren PJ, Vaes WH. Association between consumption of cruciferous vegetables and condiments and excretion in urine of isothiocyanate mercapturic acids. J Agric Food Chem 2006; 54(15): 5350-8.
[http://dx.doi.org/10.1021/jf060723n] [PMID: 16848516]

[29]  Atwell LL, Hsu A, Wong CP, *et al.* Absorption and chemopreventive targets of sulforaphane in humans following consumption of broccoli sprouts or a myrosinase-treated broccoli sprout extract. Mol Nutr Food Res 2015; 59(3): 424-33.
[http://dx.doi.org/10.1002/mnfr.201400674] [PMID: 25522265]

[30]  La Marca M, Beffy P, Della Croce C, *et al.* Structural influence of isothiocyanates on expression of cytochrome P450, phase II enzymes, and activation of Nrf2 in primary rat hepatocytes. Food Chem Toxicol 2012; 50(8): 2822-30.
[http://dx.doi.org/10.1016/j.fct.2012.05.044] [PMID: 22664424]

[31]  Wang X, He H, Lu Y, *et al.* Indole-3-carbinol inhibits tumorigenicity of hepatocellular carcinoma cells *via* suppression of microRNA-21 and upregulation of phosphatase and tensin homolog. Biochimica et Biophysica Acta (BBA) – Mol Cell Res 2015; 1853: 244-53.

[32]  Choi Y, Kim Y, Park S, Lee KW, Park T. Indole-3-carbinol prevents diet-induced obesity through modulation of multiple genes related to adipogenesis, thermogenesis or inflammation in the visceral adipose tissue of mice. J Nutr Biochem 2012; 23(12): 1732-9.
[http://dx.doi.org/10.1016/j.jnutbio.2011.12.005] [PMID: 22569347]

[33]  Crozier A, Jaganath IB, Clifford MN. Dietary phenolics: chemistry, bioavailability and effects on health. Nat Prod Rep 2009; 26(8): 1001-43.
[http://dx.doi.org/10.1039/b802662a] [PMID: 19636448]

[34]  Landete JM. Updated knowledge about polyphenols: functions, bioavailability, metabolism, and health. Crit Rev Food Sci Nutr 2012; 52(10): 936-48.
[http://dx.doi.org/10.1080/10408398.2010.513779] [PMID: 22747081]

[35]  Pereira DM, Valentao P, Pereira JA, Andrade PB. Phenolics: from chemistry to biology. Molecules 2009; 14: 2202-11.
[http://dx.doi.org/10.3390/molecules14062202]

[36]  Crozier A, Jaganath IB, Clifford MN. Phenols, polyphenols and tannins: An overview. Plant secondary metabolites: occurrence, structure and role in the human diet. Oxford, UK: Blackwell 2006; pp. 1-24.
[http://dx.doi.org/10.1002/9780470988558.ch1]

[37]  Crozier A, Jaganath IB, Clifford MN. Dietary phenolics: chemistry, bioavailability and effects on health. Nat Prod Rep 2009; 26(8): 1001-43.
[http://dx.doi.org/10.1039/b802662a] [PMID: 19636448]

[38]  Podsedek A. Natural antioxidants and antioxidant capacity of Brassica vegetables: A review. Lwt-Food Sci Technol 2007; 40: 1-11.
[http://dx.doi.org/10.1016/j.lwt.2005.07.023]

[39]    Manach C, Scalbert A, Morand C, Rémésy C, Jiménez L. Polyphenols: food sources and bioavailability. Am J Clin Nutr 2004; 79(5): 727-47.
[PMID: 15113710]

[40]    Gorinstein S, Park YS, Heo BG, *et al.* A comparative study of phenolic compounds and antioxidant and antiproliferative activities in frequently consumed raw vegetables. Eur Food Res Technol 2009; 228: 903-11.
[http://dx.doi.org/10.1007/s00217-008-1003-y]

[41]    Vallejo F, Tomas-Barberan FA, Garcia-Viguera C. Effect of climatic and sulphur fertilisation conditions, on phenolic compounds and vitamin C, in the inflorescences of eight broccoli cultivars. Eur Food Res Technol 2003; 216: 395-401.

[42]    Robbins RJ, Keck AS, Banuelos G, Finley JW. Cultivation conditions and selenium fertilization alter the phenolic profile, glucosinolate, and sulforaphane content of broccoli. J Med Food 2005; 8(2): 204-14.
[http://dx.doi.org/10.1089/jmf.2005.8.204] [PMID: 16117613]

[43]    Velasco P, Cartea ME, González C, Vilar M, Ordás A. Factors affecting the glucosinolate content of kale (*Brassica oleracea acephala* group). J Agric Food Chem 2007; 55(3): 955-62.
[http://dx.doi.org/10.1021/jf0624897] [PMID: 17263499]

[44]    Francisco M, Velasco P, Moreno D, Garcia-Viguera C, Cartea M. Cooking methods of *Brassica rapa* affect the preservation of glucosinolates, phenolics and vitamin C. Food Res Int 2010; 43: 1455-63.
[http://dx.doi.org/10.1016/j.foodres.2010.04.024]

[45]    Baardseth P, Bjerke F, Martinsen BK, Skrede G. Vitamin C, total phenolics and antioxidative activity in tip-cut green beans (*Phaseolus vulgaris*) and swede rods (*Brassica napus* var. *napobrassica*) processed by methods used in catering. J Sci Food Agric 2010; 90(7): 1245-55.
[http://dx.doi.org/10.1002/jsfa.3967] [PMID: 20394008]

[46]    Ciska E, Kozlowska H. The effect of cooking on the glucosinolates content in white cabbage. Eur Food Res Technol 2001; 212: 582-7.
[http://dx.doi.org/10.1007/s002170100293]

[47]    Czarniecka-Skubina E. Effect of the material form, storage and cooking methods on the quality of Brussels sprouts. Pol J Food Nutr Sci 2002; 11/52: 75-82.

[48]    Cartea ME, Francisco M, Soengas P, Velasco P. Phenolic compounds in Brassica vegetables. Molecules 2011; 16(1): 251-80.
[http://dx.doi.org/10.3390/molecules16010251] [PMID: 21193847]

[49]    Yokozawa T, Kim HY, Cho EJ, Choi JS, Chung HY. Antioxidant effects of isorhamnetin 3,7-di-O-beta-D-glucopyranoside isolated from mustard leaf (*Brassica juncea*) in rats with streptozotocin-induced diabetes. J Agric Food Chem 2002; 50(19): 5490-5.
[http://dx.doi.org/10.1021/jf0202133] [PMID: 12207497]

[50]    Moreno DA, Perez-Balibrea S, Ferreres F, Gil-Izquierdo A, Garcia-Viguera C. Acylated anthocyanins in broccoli sprouts. Food Chem 2010; 123: 358-63.
[http://dx.doi.org/10.1016/j.foodchem.2010.04.044]

[51]    Price KR, Casuscelli F, Colquhoun IJ, Rhodes MJ. Hydroxycinnamic acid esters from broccoli florets.

Phytochemistry 1997; 45: 1683-7.
[http://dx.doi.org/10.1016/S0031-9422(97)00246-X]

[52]    Verhagen H, de Vries A, Nijhoff WA, *et al.* Effect of Brussels sprouts on oxidative DNA-damage in man. Cancer Lett 1997; 114(1-2): 127-30.
[http://dx.doi.org/10.1016/S0304-3835(97)04641-7] [PMID: 9103270]

[53]    Zieliński H, Frias J, Piskuła M, Kozłowska H, Vidal-Valverde C. Vitamin B1 and B2, dietary fiber and minerals content of *Cruciferae* sprouts. Eur Food Res Technol 2005; 221: 78-83.
[http://dx.doi.org/10.1007/s00217-004-1119-7]

[54]    USDA. National Nutrient Database Release 20. Nutrient data Laboratory ARS. 2011. Available from: https://ndb.nal.usda.gov/

[55]    Hanson P, Yang RY, Chang LC, Ledesma L, Ledesma D. Carotenoids, ascorbic acid, minerals, and total glucosinolates in choysum (*Brassica rapa* cvg. *parachinensis*) and kailaan (*B. oleraceae* Alboglabra group) as affected by variety and wet and dry season production. J Food Compos Anal 2011; 24: 950-62.
[http://dx.doi.org/10.1016/j.jfca.2011.02.001]

[56]    Farnham MW, Grusak MA, Wang M. Calcium and magnesium concentration of inbred and hybrid broccoli heads. J Am Soc Hortic Sci 2000; 125: 344-9.

[57]    López-Berenguer C, Carvajal M, Moreno DA, García-Viguera C. Effects of microwave cooking conditions on bioactive compounds present in broccoli inflorescences. J Agric Food Chem 2007; 55(24): 10001-7.
[http://dx.doi.org/10.1021/jf071680t] [PMID: 17979232]

[58]    Domínguez-Perles R, Mena P, García-Viguera C, Moreno DA. Brassica foods as a dietary source of vitamin C: a review. Crit Rev Food Sci Nutr 2014; 54(8): 1076-91.
[http://dx.doi.org/10.1080/10408398.2011.626873] [PMID: 24499123]

[59]    Jeffery EH, Brown AF, Kurilich AC, *et al.* Variation in content of bioactive components in broccoli. J Food Compos Anal 2003; 16: 323-30.
[http://dx.doi.org/10.1016/S0889-1575(03)00045-0]

[60]    Liao K, Yin M. Individual and combined antioxidant effects of seven phenolic agents in human erythrocyte membrane ghosts and phosphatidylcholine liposome systems: importance of the partition coefficient. J Agric Food Chem 2000; 48(6): 2266-70.
[http://dx.doi.org/10.1021/jf990946w] [PMID: 10888534]

[61]    Lutsenko EA, Cárcamo JM, Golde DW. Vitamin C prevents DNA mutation induced by oxidative stress. J Biol Chem 2002; 277(19): 16895-9.
[http://dx.doi.org/10.1074/jbc.M201151200] [PMID: 11884413]

[62]    Frei B, Lawson S. Vitamin C and cancer revisited. Proc Natl Acad Sci USA 2008; 105(32): 11037-8.
[http://dx.doi.org/10.1073/pnas.0806433105] [PMID: 18682554]

[63]    Kim SY, Yoon S, Kwon SM, Park KS, Lee-Kim YC. Kale juice improves coronary artery disease risk factors in hypercholesterolemic men. Biomed Environ Sci 2008; 21(2): 91-7.
[http://dx.doi.org/10.1016/S0895-3988(08)60012-4] [PMID: 18548846]

[64]    Wright ME, Lawson KA, Weinstein SJ, *et al.* Higher baseline serum concentrations of vitamin E are

associated with lower total and cause-specific mortality in the Alpha-Tocopherol, Beta-Carotene Cancer Prevention Study. Am J Clin Nutr 2006; 84(5): 1200-7.
[PMID: 17093175]

[65]    Novotny JA, Kurilich AC, Britz SJ, Baer DJ, Clevidence BA. Vitamin K absorption and kinetics in human subjects after consumption of 13C-labelled phylloquinone from kale. Br J Nutr 2010; 104(6): 858-62.
[http://dx.doi.org/10.1017/S0007114510001182] [PMID: 20420753]

[66]    Rice-Evans CA, Sampson J, Bramley PM, Holloway DE. Why do we expect carotenoids to be antioxidants *in vivo?* Free Radic Res 1997; 26(4): 381-98.
[http://dx.doi.org/10.3109/10715769709097818] [PMID: 9167943]

[67]    Sies H, Stahl W. Vitamins E and C, beta-carotene, and other carotenoids as antioxidants. Am J Clin Nutr 1995; 62(6): 1315S-21S.
[PMID: 7495226]

[68]    Kim Y, Seo JH, Kim H. β-Carotene and lutein inhibit hydrogen peroxide-induced activation of NF-κB and IL-8 expression in gastric epithelial AGS cells. J Nutr Sci Vitaminol (Tokyo) 2011; 57(3): 216-23.
[http://dx.doi.org/10.3177/jnsv.57.216] [PMID: 21908944]

[69]    Singh J, Upadhyay AK, Prasad K, Bahadur A, Rai M. Variability of carotenes, vitamin C, E and phenolics in Brassica vegetables. J Food Compos Anal 2007; 20: 106-12.
[http://dx.doi.org/10.1016/j.jfca.2006.08.002]

[70]    Rosa E. Chemical composition. Biology of brassica coenospecies. Amsterdam: Elsevier Science BV 1999; pp. 315-57.
[http://dx.doi.org/10.1016/S0168-7972(99)80011-5]

[71]    Pasko P. Sulkowska – Ziaja K, Muszynska B, Zagrodzki P. Serotonin, melatonin, and certain indole derivatives profiles in rutabaga and kohlrabi seeds, sprouts, bulbs, and roots. LWT -. Food Sci Technol (Campinas) 2014; 59: 740-5.

[72]    Traka MH, Saha S, Huseby S, *et al.* Genetic regulation of glucoraphanin accumulation in Beneforté broccoli. New Phytol 2013; 198(4): 1085-95.
[http://dx.doi.org/10.1111/nph.12232] [PMID: 23560984]

[73]    Kumazawa Y. "Glucosinolate-contaning squeezed liquid composition derived from cruciferous plant, and process for producing same"Japan. Patent WO2010064703, 2010.

# CHAPTER 3

# Bioactive Compounds of Tomatoes as Health Promoters

**José Pinela[1,2], M. Beatriz P. P Oliveira[2], Isabel C.F.R. Ferreira[1,*]**

[1] *Mountain Research Centre (CIMO), ESA, Polytechnic Institute of Bragança, Campus de Santa Apolónia, Ap. 1172, 5301-855 Bragança, Portugal*

[2] *REQUIMTE/LAQV, Faculty of Pharmacy, University of Porto, Rua Jorge Viterbo Ferreira, n° 228, 4050-313 Porto, Portugal*

**Abstract:** Tomato (*Lycopersicon esculentum* Mill.) is one of the most consumed vegetables in the world and probably the most preferred garden crop. It is a key component of the Mediterranean diet, commonly associated with a reduced risk of chronic degenerative diseases. Currently there are a large number of tomato cultivars with different morphological and sensorial characteristics and tomato-based products, being major sources of nourishment for the world's population. Its consumption brings health benefits, linked with its high levels of bioactive ingredients. The main compounds are carotenoids such as β-carotene, a precursor of vitamin A, and mostly lycopene, which is responsible for the red colour, vitamins in particular ascorbic acid and tocopherols, phenolic compounds including hydroxycinnamic acid derivatives and flavonoids, and lectins. The content of these compounds is variety dependent. Besides, unlike unripe tomatoes, which contain a high content of tomatine (glycoalkaloid) but no lycopene, ripe red tomatoes contain high amounts of lycopene and a lower quantity of glycoalkaloids. Current studies demonstrate the several benefits of these bioactive compounds, either isolated or in combined extracts, namely anticarcinogenic, cardioprotective and hepatoprotective effects among other health benefits, mainly due to its antioxidant and anti-inflammatory properties. The chemistry, bioavailability and bioactivity of these bioactive compounds will be discussed, as well as the main mechanisms of action against cancer and other bioactivities including antioxidant, anti-inflammatory, cardiovascular and hepatoprotective effects in humans. Possible applications of tomato bioactive compounds in the industry will also be proposed.

*  **Corresponding author Isabel C.F.R. Ferreira:** Mountain Research Centre (CIMO), ESA, Polytechnic Institute of Bragança,  Campus de Santa Apolónia,  Apartado 1172, 5301-855 Bragança, Portugal; Tel.: + 351 273303219; Fax: + 351 273325405; Email: iferreira@ipb.pt.

**Luís Rodrigues da Silva and Branca Maria Silva (Eds.)**

**Keywords:** Anticarcinogenic, Anti-inflammatory, Antioxidant, Ascorbic Acid, Bioactivity, Bioavailability, Cardioprotection, Clinical trials, Flavonoids, Functional food, Glycoalkaloids, Human health, Hydroxycinnamic acids, Lycopene, *Lycopersicon esculentum*, Nutraceuticals, Phenolic compounds, Tocopherols, Tomatine, Tomato.

## INTRODUCTION

The tomato plant (*Lycopersicon esculentum* Mill.) was imported from the Andean region to Europe in the 16th century. It belongs to the Solanaceae family that includes many other plants of economic importance, including potatoes, eggplants, peppers and tobacco [1]. Today, this species is widespread throughout the world, representing the most economically important vegetable crop worldwide. In fact, tomato is the most consumed vegetable after potatoes and probably the most preferred garden crop. In 2013, about 164 million tonnes of tomatoes were produced in the world, having been registered an increase above 2.6 million tonnes over 2012. The three main producing countries are China, India and United States of America, but it is in the Mediterranean and Arabian countries that their consumption is higher [2].

Tomato is a very versatile fruit, being consumed fresh but also processed as paste, soup, juice, sauce, powder, or concentrate. In addition, there are several tomato cultivars and varieties with a wide range of morphological and sensorial characteristics which affect the way how they are prepared and consumed [1, 3, 4]. Tomatoes and tomato-based food products are an important source of nourishment for the world's population. Regarding its nutritional value, if one takes into consideration only the proteins, fat, carbohydrates, or sugars content, it appears clearly that it does not have a high nutritional value. However, it represents an important source of other nutrients and non-nutrients endowed with important health promoting properties, namely carotenoids such as β-carotene (provitamin A) and mostly lycopene, which provides the deep red colour, vitamins such as ascorbic acid (vitamin C) and tocopherols (vitamin E), phenolic compounds including hydroxycinnamic acid derivatives and flavonoids, lectins, and minerals (K, Mn, Ca, Cu and Zn) [3 - 5].

Tomatoes are the most important component of the Mediterranean diet, known to be beneficial for human health [6]. A relationship between the consumption of tomatoes and tomato-based foods and the prevention of chronic degenerative disease induced by oxidative stress and inflammation has been indicated in several studies [7 - 10]. However, the bioaccessibility and bioavailability of tomato compounds is affected by the way how tomatoes are consumed (*i.e.*, raw or processed), which affects its subsequent bioactivity. Clinical trials performed in the last years elucidate the positive effects and mechanisms involved in the activity of tomato compounds against cardiovascular disease and various types of cancer [9 - 12]. Indeed, tomato extracts, as well as lycopene, α-tomatine and some phenolic compounds have been highlighted as having increased potential for the development of new drugs, nutraceuticals and functional foods.

This chapter highlights the tomato fruit as a functional food and as a source of nutraceutical ingredients of industrial value. In this sense, the major health promoting compounds of tomatoes are described chemically and their bio-availability, bioactivity and impact on human health are discussed. Recent *in vitro* and *in vivo* clinical trials are presented, with particular attention paid to the mechanisms behind the protective effects of tomato bioactive compounds against the most common degenerative diseases associated to oxidative stress and inflammation, including cardiovascular and hepatocellular diseases, diabetes and various types of cancer, among other health problems.

## TOMATO BIOACTIVE COMPOUNDS

Nowadays, consumers are increasingly made aware and better informed about the health benefits provided by food beyond its basic nutritional role. Actually, they are looking for foods with health promoting properties called "functional foods". The tomato fruit is a good example, whose functionality or health claim properties are conferred by biologically active ingredients responsible for decreasing the risk of susceptibility to certain diseases. The major compounds of this fruit (Fig. **1**) are carotenoids (β-carotene and mainly lycopene), vitamins (ascorbic acid and tocopherols), and phenolic compounds including hydrocinnamic acids (mainly caffeic acid and its ester chlorogenic acid) and flavonoids such as narigenin and rutin [4, 5, 13]. Other bioactive compounds, such as glycoalkaloids and lectins also present in tomato fruits have shown relevant biological effects *in vitro* and *in vivo* [14, 15]. Nevertheless, the content of bioactive compounds in tomatoes is affected by environmental and genetic (cultivar or variety) factors, geographical location, agricultural practices, processing conditions, among others [4, 16 - 20]. The main bioactive compounds of tomato fruit are described below.

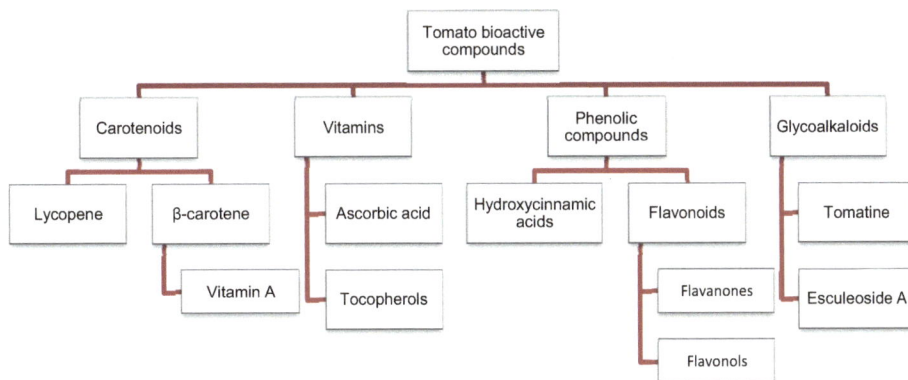

**Fig. (1).** Groups of bioactive compounds of tomatoes.

## Carotenoids

Carotenoids are a class of hydrocarbons consisting of eight isoprene units, which are joined in a head-to-tail pattern (except at the centre) to confer symmetry to the molecular structure. This way, the two central methyl groups are in a 1,6-positional relationship and the remaining non-terminal methyl groups are in a 1,5-positional relationship. Most carotenoids derive from a 40-carbon polyene chain structure, which is considered as the backbone of these compounds. This chain can terminate with cyclic end-groups (rings) and may have oxygen-containing functional groups. The long unsaturated alkyl chains make carotenoids highly lipophilic molecules. In higher plants, carotenoids can be found in chloroplasts of photosynthetic tissues and in chromoplasts of flowers and fruits. Generally, they occur in the free form in leaves and in the esterified form in other tissues. These natural pigments play a central role in photosynthesis; they are involved in photosystem assembly and light harvesting, and protect from excessive light through energy dissipation and free radical elimination, thereby reducing membrane damage. In humans, carotenoids are part of the antioxidant defence system and interact synergistically with other bioactive compounds [21]. Once animals cannot synthesize carotenoids, they need to be incorporated through diet, being tomatoes and tomato based-products one of the most common sources of carotenoids available for the human population.

Tomatoes and tomato-based foods account for over 85% of all the dietary sources of lycopene [22]. Lycopene is the most abundant carotenoid in ripe tomatoes, representing about 80 to 90% of these pigments [23]. Normally, tomatoes contain up to 10 mg lycopene per 100 g of fresh weight [3], depending on some factors such as those mentioned above. Additionally, the lycopene content increase as the fruit ripens [24]. Chemically, lycopene (Fig. **2**) is a polyunsaturated (polyene) straight-chain molecule with 11 conjugated and 2 nonconjugated double bonds. Thus, it can be found in both the *cis* and *trans* configurations because of the presence of the double bonds [25]. Additionally, its straight structure facilitates its incorporation into some organs such as the liver, adrenal glands and prostate, where it has a role of preventing oxidative reactions associated with the outbreak of different diseases [26]. The *trans* configuration is the most common isomer and largely responsible for the deep red colour of the ripe red tomatoes [25, 27]. Nevertheless, the *trans* form is prone to isomerisation under the influence of some processing conditions, which include the action of light, heat, oxygen and acids, and, after ingestion, it is partly transformed *in vivo* to the more bioactive *cis* form [14, 28]. Lycopene has strong antioxidant activity and other *in vitro* and *in vivo* beneficial effects because of its capacity to act as a free radical scavenger [29] that is twice that of β-carotene (Fig. **2**) [22]. In ripe red tomato fruits, the ratio of lycopene to β-carotene content varies widely between 1.5 and 40 [30, 31]. Along with α-carotene and β-cryptoxanthin, β-carotene is a provitamin A carotenoid, *i.e.*, it can be converted by the human body into two molecules of vitamin A (Fig. **2**). Actually, what we generally call vitamin A is a group of naturally-occurring molecules, structurally similar to retinol, that are capable of exerting biological activity [32]. In addition to lycopene and β-carotene, phytoene, phytofluene, ζ-carotene, γ-carotene, neurosporene and lutein are other carotenoids reported in tomatoes and tomato-based products [33].

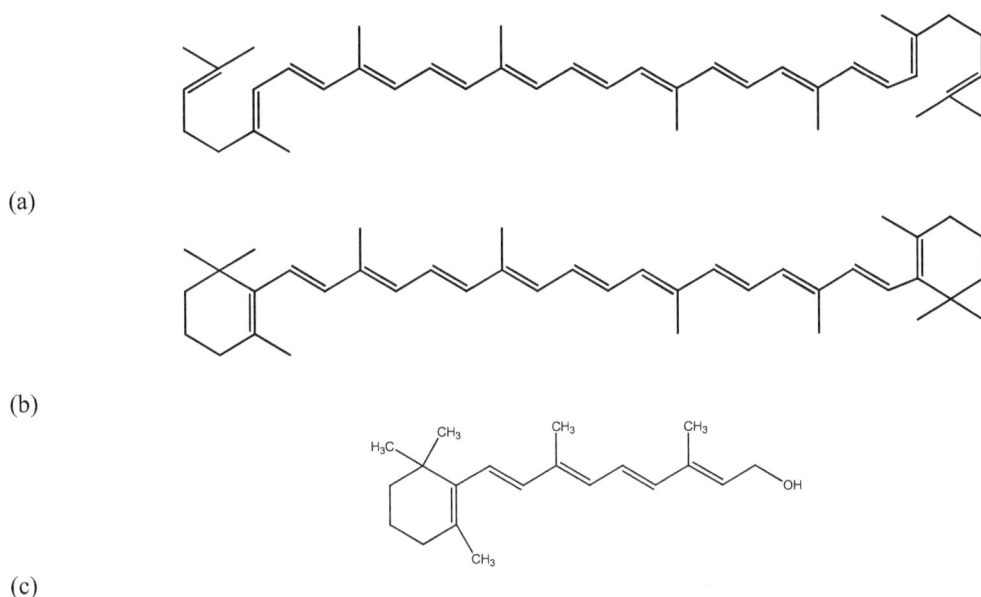

(a)

(b)

(c)

**Fig. (2).** Chemical structures of **(a)** lycopene, **(b)** β-carotene and **(c)** vitamin A.

## Vitamins

Vitamin C and E are the respective generic names for ascorbic acid and tocopherols. Ascorbic acid, a 6-carbon lactone ring structure with a 2,3-enediol moiety, can be found in all living and actively metabolising plant parts and cell compartments [34]. It comprises two compounds endowed with bioactivity: $_L$-ascorbic acid and $_L$-dehydroascorbic acid (Fig. **3**). Both are easily absorbed by the gastrointestinal tract and can interchange enzymatically *in vivo*. In biological systems, ascorbic acid exists as the monovalent anion $_L$-ascorbate [35]. However, this vitamin is highly susceptible to oxidation in the presence of metal ions like $Cu^{2+}$ and $Fe^{3+}$. Its oxidation is also influenced by light, heat, pH, oxygen and water activity [36]. This vitamin has the ability to act as electron donor, being a potent *in vivo* antioxidant; it protects low-density lipoproteins (LDL) from the oxidation caused by different oxidative stress reactions and inhibits the LDL oxidation caused by vascular endothelial cells. The high volume of consumption of tomato all the year round makes this fruit one of the main sources for this vitamin.

Different levels of ascorbic acid have been reported in tomatoes (8-21 mg/100 g of fresh weight) [4, 37, 38], since it is affected by different factors. In turn, vitamin E includes eight chemically distinct molecules (Fig. **4**), four tocopherol isoforms (α-, β-, γ- and δ-tocopherol) and four tocotrienol isoforms (α-, β-, δ- and γ-tocotrienol) [39]. Tocopherols differ from the corresponding tocotrienols in the aliphatic tail; tocopherols have a phytyl side chain attached to the chromanol head, whereas the tail of tocotrienols contains three *trans* double bonds at the 3', 7' and 11' positions and forms an isoprenoid chain. These unsaturations in the tail of tocotrienols give only a single stereoisomeric carbon, whereas tocopherols have eight possible stereoisomers per structural formula. The various isoforms differ in the methyl substituents on the chromanol head; the α-forms contain three methyl groups, the β- and γ- have two and the δ-forms have only one methyl group. Together, tocopherols and tocotrienols are called tocochromanols. All these compounds feature a chromanol ring with a hydroxyl group capable of donating a hydrogen atom, and a lipophilic side chain that allows for penetration into cell membranes [34]. The donation of hydrogen atoms to the peroxyl radicals forms unreactive tocopheroxyl radicals (TO˙) unable to continue the oxidative chain reaction [40]. The human body absorbs all forms of vitamin E, but maintains only the α-tocopherol [41]. The amounts of tocopherols also vary in tomatoes, having been reported values from 0.17 to 1.44 mg/100 g of fresh weight [4, 38]. Nevertheless, neither vitamin C nor vitamin E can be synthesized by humans, so their intake must be guarantee through the diet [34, 42].

(a)          (b)

**Fig. (3).** Chemical structures of **(a)** ∟-ascorbic acid and **(b)** ∟-dehydroascorbic acid.

The tomato fruit also presents folates (12-18 µg/100 g of fresh weight) [43, 44], a complex group of water-soluble compounds known as vitamin $B_9$. Folic acid (Fig. **5**) consists in an aromatic pteridine ring attached by a methylene bridge to a residue of *p*-aminobenzoic acid which, in turn, is joined by an amide bond to a glutamic acid residue [45]. Folate vitamers are involved in multiple physiological mechanisms in the field of andrology and gynecology [46] and are essential for fetal growth [47]. Folates also are involved in the homocysteine metabolism regulation [48] and some authors consider the hyperhomocysteinemia condition as a marker or risk factor for cardiovascular diseases [49, 50]. As folates are synthesised just by plants and microorganisms, humans are dependent on dietary sources like the tomato fruit for the intake of this vitamin.

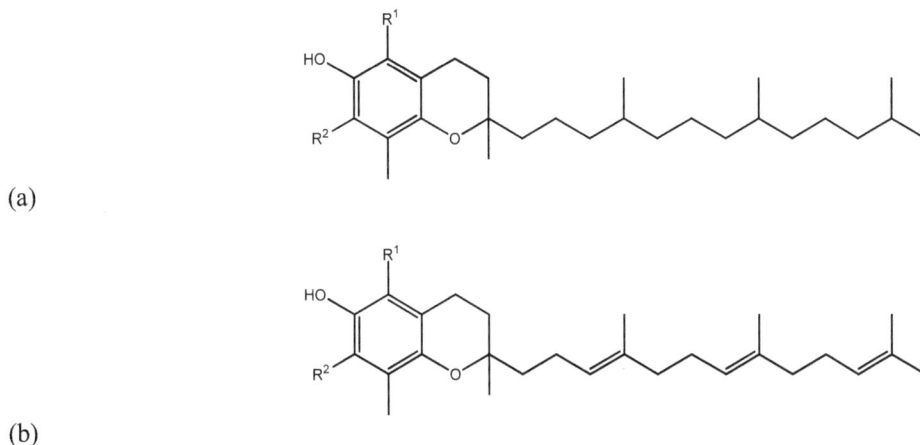

(a)

(b)

**Fig. (4).** Chemical structures of **(a)** tocopherols ($R^1 = R^2 =$ Me: α-tocopherol; $R^1 =$ Me, $R^2 =$ H: β-tocopherol; $R^1 =$ H, $R^2 =$ Me: γ-tocopherol; $R^1 = R^2 =$ H: δ-tocopherol) and **(b)** tocotrienols ($R^1 = R^2 =$ Me: α-tocotrienol; $R^1 =$ Me, $R^2 =$ H: β-tocotrienol; $R^1 =$ H, $R^2 =$ Me: γ-tocotrienol; $R^1 = R^2 =$ H: δ-tocotrienol).

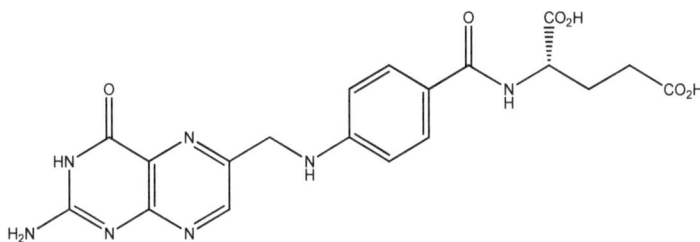

**Fig. (5).** Chemical structure of folic acid.

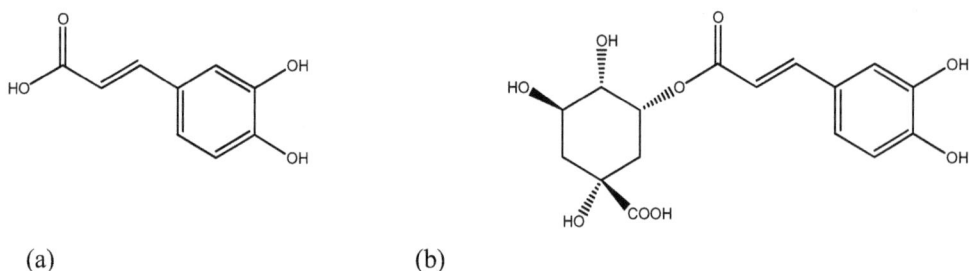

(a)                                                                                  (b)

**Fig. (6).** Chemical structures of **(a)** caffeic acid and **(b)** chlorogenic acid.

## Phenolic Compounds

Phenolic compounds are broadly spread throughout the plant kingdom, representing more than 8000 different phenolic structures. They have at least one aromatic ring with one or more hydroxyl groups attached and vary from low molecular weight molecules to large and complex ones. Phenolic compounds generally appear as esters and glycosides rather than as free compounds due to the conferred stability of these molecules. Phenolic (hydroxycinnamic) acids and flavonoids are the most abundant phenolic compounds in tomatoes [5], as well as in the diet [51]. The phenolic acids represent a group of compounds that derive from cinnamic acid through the phenilpropanoid pathway. They display a $C_6$-$C_3$ skeleton of *trans*-phenyl-3-propenoic acid with one or more hydroxyl groups bonded to the phenyl moiety, some of which may be methylated. According to literature, caffeic acid and its ester chlorogenic acid are the main phenolic acids in tomato (Fig. **6**) and the most extensively studied [5, 52]. Both compounds have *in vitro* antioxidant activity [53] and might inhibit the formation of mutagenic and carcinogenic *N*-nitroso compounds [54]. Curiously, the antioxidant mechanism of chlorogenic acid is analogous to that of lycopene. The flavonoids are the largest group of molecules within the phenolic compounds. These polyphenolic compounds display an arrangement of three aromatic rings with 15 carbons and a $C_6$-$C_3$-$C_6$ skeleton which can have numerous substituents. The A ring is a benzene, condensed with a six-member ring (ring C), which carries a phenyl benzene in position 2 (ring B) as a substituent. Sugars in the form of glycosides are normally joined with flavonoids and while they increase water solubility along with hydroxyl groups, other substituents like methyl groups and isopentenyl units increase their lipophilic properties [39]. In tomatoes, flavonoids are represented by flavanones, including naringenin glycosylated derivatives, and flavonols such as quercetin, rutin and kaempferol glycosylated derivatives (Fig. **7**) [55 - 58]. They are commonly found in the skin and in small amount in the other parts of the fruit.

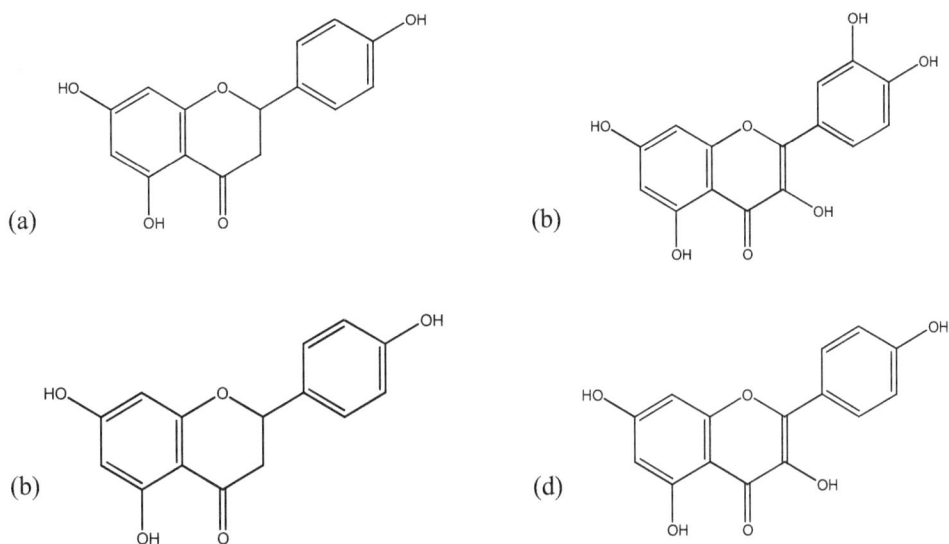

**Fig. (7).** Chemical structures of (**a**) naringenin, (**b**) quercetin, (**c**) rutin and (**d**) kaempferol.

## Glycoalkaloids

Glycoalkaloids are characteristic secondary metabolites in plants of the Solanaceae family. They are involved in host-plant resistance and have pharmacological and nutritional effects in humans and animals. In tomato plants, the glycoalkaloids tomatine and esculeoside A (Fig. **8**) are synthesized. Tomatine comprises a junction of α-tomatine and dehydrotomatine (Fig. **8**). Structurally, dehydrotomatine differs from α-tomatine by having a double bond in the steroidal ring B of the aglycone. However, both glycoalkaloids have the same tetra-saccharide (lycotetraose) side chain; α-tomatine has lycotetraose bonded to the aglycone tomatidine, whereas dehydrotomatine has the side chain attached to an aglycone tomatidenol [14, 15]. Unripe tomatoes may contain up to 500 mg of tomatine per kg of fresh weight; but the levels are decreased with the ripening of tomatoes and, therefore, ripe red tomatoes present lower levels (~5 mg/kg of fresh weight) [59].

(a)

(b)

(c)

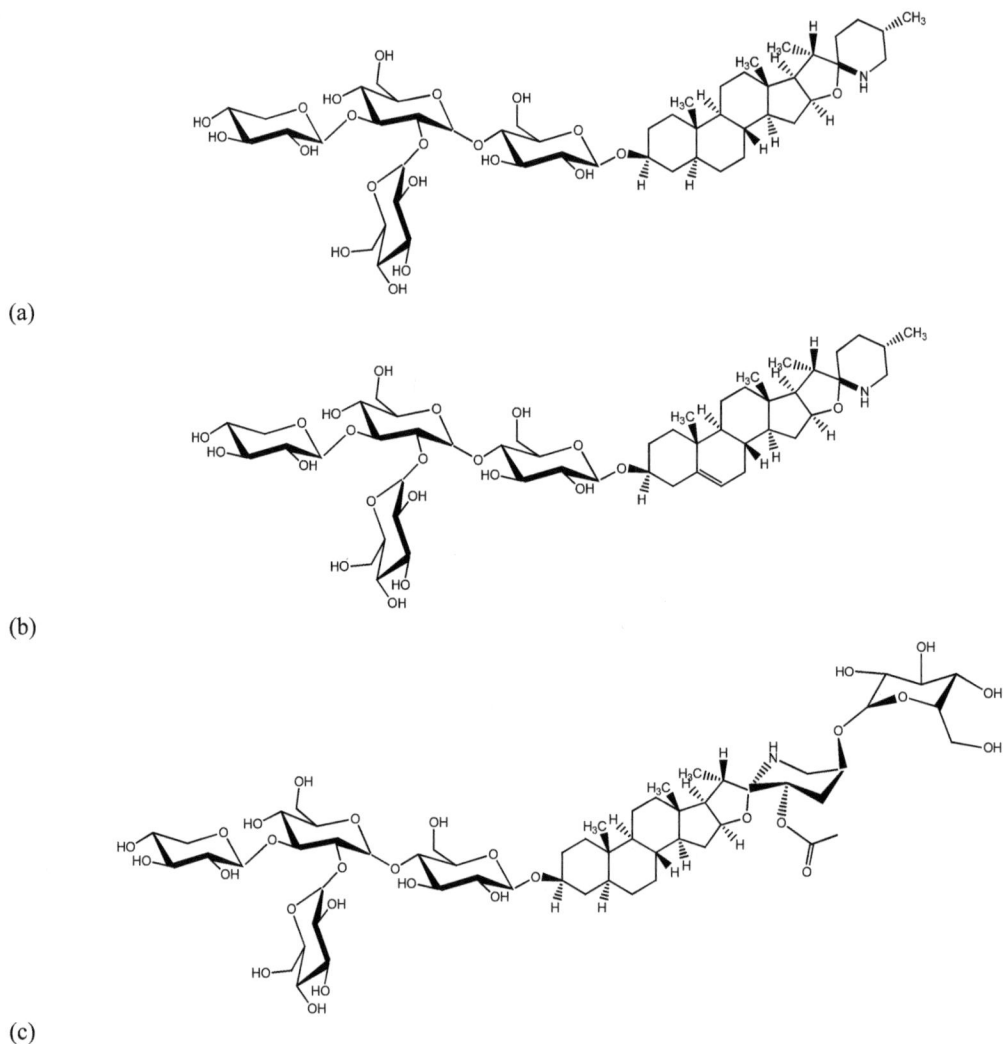

**Fig. (8).** Chemical structures of **(a)** α-tomatine, **(b)** dehydrotomatine and **(c)** esculeoside A.

Beside, the tomatine content of cherry tomatoes is several fold greater than that of larger size standard varieties. On the other hand, the content of esculeoside A, which is stored in ripe fruits of cultivated tomatoes, is comparable to or higher than that of lycopene [60, 61]. Thus, the levels of esculeoside A increase as the fruit matures, contrary to that observed for tomatine [14, 62].

## BIOAVAILABILITY OF TOMATO COMPOUNDS

The bioavailability of bioactive compounds is crucial to their physiologic effect. Before becoming bioavailable, the tomato bioactive compounds must be released from the plant matrix and modified in the gastrointestinal tract. The digestive transformations that involve the conversions of tomato into substances ready to be absorbed and assimilated are called bioaccessibility. It is commonly assessed using *in vitro* digestion assays, which simulate the gastric and small intestinal digestion processes. Differently, the term bioavailability can be defined as the fraction of a compound or metabolite that reaches the systemic circulation. It is evaluated using *in vivo* assays in animals or humans by measuring the concentration of a compound in plasma or urine after administration of an acute or chronic dose of the isolated compound or compound-containing food [63].

### Bioavailability of Carotenoids

The bioavailability of the tomato carotenoids is widely affected by endogenous (tomato-related) and exogenous (processing-related) factors. Firstly, in order to become bioaccessible, carotenoids need to be released from the tomato matrix in which they are embedded. Thereafter, they need to solubilise into an oil phase either during processing and/or during the gastric digestion. The release of carotenoids from the tomato matrix and its subsequent incorporation in the oil and micellar phase are decisive steps during digestion in order to become bioavailable.

In fact, the mixed micelles formed in the small intestine are the primary vehicle for the absorption of carotenoids *via* intestinal mucosa [64]. That's why the bioavailability is greatly affected by dietary composition, *i.e.*, the co-ingestion of carotenoids and fat is very important and necessary for absorption [65 - 67]. Actually, the Mediterranean way of preparing tomatoes, by cooking them in olive oil, is a smart way to promote the absorption of these bioactive compounds. Regarding exogenous factors, it is know that the lycopene bioaccessibility increases under processing conditions because cell membranes are disrupted, which increases its release from the tomato matrix [21]. Thermal treatments also promote the lycopene isomerisation of *trans* to *cis* form, isomer that is described as being more bioavailable [28, 68]. Thus, lycopene from tomato-based processed foods is generally more bioavailable than from fresh tomatoes. Nevertheless, inadequate processing and storage conditions can cause isomerisation during the formation of by-products, which can reduce the absorption of carotenoids and make the food less desirable to the consumer.

## Bioavailability of Vitamins

The ingestion of ascorbic acid causes a dose-dependent increase of this vitamin in the plasma. Its absorption from the gastrointestinal tract occurs by a sodium-dependent active transport mechanism (mainly in the jejunum) but also by a passive absorptive pathway. The active transport predominates when ascorbic acid is at low gastrointestinal concentrations, but when at high concentrations the active transport becomes saturated allowing only passive diffusion [35]. In turn, the absorption of tocopherols is similar to that observed for other fat-soluble vitamins, being necessary it's packaging into micelles (emulsified in the presence of bile salts and amphipathic lipids available in the intestinal lumen) to facilitate their absorption during digestion, the same as carotenoids. After entry into the intestinal absorptive cells (enterocytes), tocopherols are packaged into chylomicrons and enter the circulation through the lymph-vascular system. Thereafter, chylomicrons triglycerides are hydrolyzed by endothelial bound lipoprotein lipases, resulting in the transfer of tocopherols and lipids to peripheral tissues [69]. However, the main steps and molecular mediators of the tocopherols transport from the luminal micellar phase into the enterocytes are not yet fully elucidated.

## Bioavailability of Phenolic Compounds

The bioavailability within the class of phenolic compounds is widely variable and the most abundant in our diet do not always correspond to those with better bioavailability. The ability of the human body to absorb and metabolize these compounds varies widely depending primarily on their physicochemical properties, such as the basic structure, molecular size, degree of polymerization or glycosylation, solubility and conjugation with other phenolics [63]. Phenolic acids of low-molecular weight such as gallic acid are easily absorbed by the small intestine, as well as flavones and quercetin glycosides [70]. Conversely, larger polyphenols are poorly absorbed. In addition, a large number of phenolic compounds is found in the glycosylated form or as esters or polymers which must be hydrolyzed before the free aglycone can be absorbed. The human metabolism also greatly affects the bioavailability of these bioactive compounds. Once absorbed, polyphenols undergo biotransformations in the small intestine, and later in the liver, into various *O*-sulfated, *O*-glucuronidated and *O*-methylated forms. Thus, the chemical structure of the resultant metabolites could be quite different from that of the parent compounds and, therefore, these metabolites may or not have the initial biological activity [71]. Moreover, the bioavailability of the phenolic compounds can be influenced by different food processing steps. For example, it has been demonstrated a significant increase in the levels of chlorogenic acid and naringenin in plasma after consumption of cooked tomatoes in comparison with fresh tomatoes [72]. Actually, mechanical and thermal treatments involved in the manufacture of tomato sauce helps to release these bioactive compounds from the tomato matrix, thus increasing the bioavailability more efficiently than through the addition of oil [73].

### Bioavailability of Glycoalkaloids

Very limited studies have been conducted to study the bioavailability of glycoalkaloids in humans, but it is known that these compounds are poorly absorbed by the gastrointestinal tract of mammals. An appreciable amount of the ingested glycoalkaloids is hydrolyzed in the gut to less toxic aglycones, and the originated metabolites are rapidly excreted *via* urine and feces [74].

### BIOACTIVITY OF TOMATO COMPOUNDS

Bioactivity can be defined as the effect caused upon exposure to an active ingredient. It comprises tissue uptake, as previously referred, and the resulting physiological response, which can be evaluated *in vivo*, *ex vivo* or using *in vitro* assays. The tomato compounds are known for their capacity to act as free radical scavengers of reactive oxygen and nitrogen species (ROS/RNS). These species include free radicals and other non-radical reactive substances also called oxidants. ROS and RNS are generated as a normal part of human metabolism and its production can be promoted by external factors. The accumulation of these species in the body gives rise to a phenomenon known as oxidative stress, which results from an imbalance between generation and neutralization of reactive species in the cells. The main targets of these species are proteins, deoxyribonucleic acid (DNA) and ribonucleic acid (RNA) molecules, lipids and sugars. The lipid peroxidation is one of the most undesirable effects of ROS because of the consequent formation of free radicals. This phenomenon is initiated by an attack towards a fatty acid side chain by a radical in order to abstract a hydrogen atom. A higher number of double bonds in the fatty acid facilitate the removal of hydrogen atoms and consequently the formation of radicals. After that, the carbon-centred lipid radical can undergo molecular rearrangement and react with oxygen forming a peroxyl radical. These highly reactive species can abstract hydrogen atoms from surrounding molecules and propagate the lipid peroxidation chain reaction. Hydroxyl radicals are major radicals in lipid peroxidation mechanisms [39, 75]. Products generated by these chain reactions are toxic, *e.g.*, malondialdehyde (MDA) may be involved in the onset of mutagenic damage. ROS can also activate the transcription nuclear factor kappa B (NF-κB), which leads to the expression of pro- and anti-inflammatory cytokines genes and their subsequent production [76]. As a consequence, these processes play a key role in the development of several degenerative and chronic diseases, as well in aging. The main mechanisms and properties inherent to the bioactivity and protective effects of the major tomato compounds are discussed below.

## Bioactive Properties of Carotenoids

Carotenoids can inhibit the lipid peroxidation due to their capacity to act as free radical scavengers [29]. The basic antioxidant properties of these pigments are conferred by its singlet oxygen quenching capacity, by which the carotenoids are excited. The excited carotenoids can dissipate the excess energy through a sequence of rotational and vibracional interactions that allows them to return to the unexcited state, and so quench more radicals. Indeed, these bioactive compounds are known for its capacity to scavenge peroxyl radicals more efficiently as compared to others ROS [77]; however, these radicals are the only ones able to annihilate these pigments [39]. Carotenoids may also decay and form non-radical compounds able to stop free radical attacks through its binding to these radical species [39, 78]. Nevertheless, its effects in humans are quite complex and it is still unclear whether these biological effects result from their antioxidant capacity or of a non-antioxidant mechanism [34]. According to Navarro-González *et al.* [27], the mechanism of action of lycopene include: its role as an antioxidant, decreasing the LDL oxidation and the lipid peroxidation and lowering the LDL cholesterol and the total cholesterol, and as a modulator of inflammatory response through the reduction of cytokines implicated in cardiovascular disease. Lycopene has also influence on cellular proliferation and differentiation as well as in the immune response [79]. Curiously, lycopene has the capacity to inhibit the lipopolysaccharide-induced phenotypic and functional maturation of murine dendritic cells both *in vitro* and *in vivo* [80]. It also reduces the oxidative stress and intestinal inflammation in experimental models of colitis in rats [81]. As mentioned above, the high number of conjugated double bonds in lycopene structure provides the singlet oxygen quenching capacity. Lycopene and β-carotene reduce the production of LDL cholesterol oxidized products that are associated with coronary heart disease; β-carotene also protects the skin against deleterious effects of sunlight [77]. The antioxidant potential of carotenoids is commonly linked to its capacity to prevent free radical triggered diseases, including atherosclerosis, multiple sclerosis, age-related muscular degeneration and cataracts [36]. In fact, the consumption of tomatoes and tomato-based foods has been significantly connected to a low incidence of prostate cancer [82].

## Bioactive Properties of Vitamins

The antioxidant activity of tocopherols is conferred by the chromanol head group. The phytyl side chain has no activity; it is embedded within the cell membrane bilayer while the active chromanol ring is closely positioned to the surface. This ingenious arrangement allows tocopherols to act as powerful antioxidants and to be regenerated through reaction with other antioxidants, *e.g.*, ascorbic acid [36]. However, the activity of these antioxidants is affected by its orientation within the membrane. Thus, the tocopherols halts lipid peroxidation in cell membranes and various lipid particles through donation of the phenolic hydrogen of the chromanol ring to lipid peroxyl radicals, thereby forming unreactive tocopheroxyl radicals unable to continue the oxidative chain reaction [83]. This vitamin protects LDL and cell membrane polyunsaturated fatty acids, and inhibits smooth muscle cell proliferation and protein kinase C activity [36]. Curiously, tocopherols are the major lipid-soluble antioxidants found in plasma, red cells and tissues [40]. They have been associated with lower incidence of heart disease, delay of Alzheimer's disease, and prevention of several types of cancer. The α-tocopherol, for example, can reduce the nitrogen dioxide levels more effectively than the other isoforms, a compound implicated in arthritis and carcinogenic processes [36]. However, tocopherols are not efficient scavengers of hydroxyl radicals *in vivo* [84].

The bioactivity of ascorbic acid is conferred by the 2,3-enediol. The antioxidant mechanisms are conferred by its ability to donate a hydrogen atom to free radicals, to eliminate molecular oxygen and quench singlet oxygen. The capacity to scavenge aqueous radicals and regenerate α-tocopherol from tocopheroxyl radicals are other well-known antioxidant mechanisms of this vitamin [36, 84]. In biological systems, ascorbic acid changes to the ascorbate radical through the donation of an electron to a lipid radical in order to stop the lipid peroxidation chain reaction. Thereafter, the originated ascorbate radicals react rapidly and produce one molecule of ascorbate and other of dehydroascorbate. The last one is devoid of bioactivity, but is converted back into ascorbate [77]. Ascorbic acid is efficient in scavenging the superoxide radical anion, hydroxyl radicals, singlet oxygen, hydrogen peroxide and reactive nitrogen oxide. However, the ascorbic acid may also act as a prooxidant, for example during the reduction of ferric iron ($Fe^{3+}$) to the more active ferrous iron ($Fe^{2+}$) [36].

The vitamin A has high antioxidant activity and can protect lipids against rancidity. In humans, it plays a vital role in protecting LDL from oxidation stimulated by copper [85]. Retinoids are essential to diverse physiological functions including vision, immune response, bone mineralization, reproduction, cell differentiation, and growth [32].

**Bioactive Properties of Phenolic Compounds**

The antioxidant capacity of phenolic acids comes from its ability to chelate transition metals and to scavenge free radicals, having a significant impact over hydroxyl and peroxyl radicals, peroxynitrites and superoxide anions [39]. The hydroxycinnamic acid derivatives display bioactivity due to the hydroxylation and methylation patterns of the aromatic ring, *e.g.*, the *o*-dihydroxy group in the phenolic ring of caffeic acid improves its antioxidant capacity [86]. The free radical scavenging mechanism of the hydroxycinnamic acids is analogous to that of flavanoids, which is attributed to its capacity to donate a hydroxyl hydrogen and resonance stabilization of the resulting radicals. The *o*-dihydroxy substituents also have iron-chelating ability [77].

The bioactivity of flavonoids is conferred by hydroxyl groups attached to the ring structures. They can act as reducing agents, hydrogen donators, superoxide radical scavengers, singlet oxygen quenchers, and metal chelators. Flavonoids have capacity to protect the DNA from damage caused by hydroxyl radicals, reduce tocopheroxyl radicals, activate or inhibit bioactive enzymes, and alleviate the nitrosative stress [39, 77, 87]. Rutin, also known as vitamin P, has antioxidant, anti-inflammatory, and anticarcinogenic properties. It reduces the fragility of blood vessels [88]. At a low concentration of cupric ions, quercetin is capable of protecting DNA from oxidative damage resulting from the attack of the hydroxyl and superoxide radicals and hydrogen peroxide [77].

## Bioactive Properties of Glycoalkaloids

Glycoalkaloids are perceived as potentially toxic, but the studies conducted over the past decade indicate that they may also have health-promoting effects, depending on dose and conditions of use. They can be used because of its anti-inflammatory, anticancer, antipyretic, anticholesterol, antinociceptive, and anti-microbial effects [89]. Its bioactivity derives mainly from the capacity to inhibit acetylcholinesterase (AChE) and butyrylcholinesterase (BuChE), and from its ability to complex with membrane 3$\beta$-hydroxy sterols, which causes the rupture of the membrane. Its capacity to inhibit AChE and BuChE makes them important compounds, but the source of enzymes to be tested (plasma *vs.* serum) and differences in the purity of the glycoalkaloid and aglycones have affected the published results [89, 90]. However, the aglycone alone is practically inactive against the cholinesterase enzymes. The sugar unit is required for activity, but it is the structure of the aglycone which determines the extent of inhibition. The existence of heterocyclic nitrogen is also a necessary condition for activity [91]. Regarding the second mechanism of action, and with respect to the aglycone subunit, an intact E ring and an unshared pair of electrons on the nitrogen of the F ring, as well as solanidane and spirosalane rings are necessary for membrane lytic activity [92]. In general, the glycoalkaloids bioactivity increases when they are administered as mixtures (depending on the relative proportion used) [89]. However, the synergistic activity of α-tomatine and dehydrotomatine remains unknown. In inflammatory processes, the aglycone tomatidine has the capacity to reduce inducible nitric oxide synthase (iNOS) and cyclooxygenase-2 (COX-2) expression through blocking NF-κB and JNK signalling in lipopolysaccharide-stimulated macrophages [93]. In turn, α-tomatine has the ability to decrease the cholesterol and triglycerols levels, enhance the immune system, and inhibit the growth of cancer cells [94 - 96]. Actually, this compound has been recognized as a potential anticancer drug [97]. Nevertheless, a deeper understanding of the implications of glycoalkaloids in the human diet is still necessary.

## TOMATO AND HUMAN HEALTH

The regular consumption of tomatoes and tomato-based foods has been associated with several positive effects on human health. Current studies demonstrate the several benefits of tomato bioactive compounds, either isolated or in combined extracts, namely anticarcinogenic [9, 82], cardioprotective [8, 12, 98], anticholes-terolemic [99, 100], antidiabetic [101], and hepatoprotective [102, 103] effects among other health benefits, mainly due to its antioxidant [4, 8, 104] and anti-inflammatory [7, 104] properties. Indeed, the production of ROS and RNS during oxidative stress and inflammatory processes is widely associated with the development and progression of chronic diseases such as cancers, cardiovascular diseases, diabetes, and other disorders associated with aging [105, 106]. The tomato bioactive compounds can neutralize the generated reactive species and, therefore, prevent the associated diseases. In fact, low levels of antioxidants have been associated with heart diseases and different types of cancer [107]. Examples of *in vitro* and *in vivo* clinical trials highlighting the crucial role of tomato and tomato-derived compounds on human health are presented below.

## Tomato Consumption Improves the Oxidative Status

The involvement of oxidative stress in the pathogenesis of various degenerative diseases is evidenced by an altered expression of enzymatic antioxidant defences, such as superoxide dismutase (SOD), catalase (CAT), and glutathione peroxidase (GPx) [108]. A recent study conducted by Li *et al.* [104] demonstrated that the anti-inflammatory effect of the purple tomato extract might be caused by the decreased levels of MDA and nitric oxide (NO) and increased GPx and SOD activity in oedematous tissue. These results were attributed to the direct antioxidant activities of the bioactive compounds towards the free radicals, and indirect elevation of the enzyme activity. Supplementation of ultra marathon runners for a period of two months with a whey protein bar and tomato juice also improved the oxidative status, decreasing thiobarbituric acid reactive substances (TBARS) and protein carbonyls [8]. Other study evaluated the effect of tomato sauces with different amounts of lycopene on oxidative stress biomarkers [109]. Healthy participants consumed 160 g/day of tomato sauce, while maintaining constant their usual diet and physical activity. The regular consumption of the lycopene-enriched tomato sauce caused a considerable decrease of the oxidized-LDL cholesterol levels and increased the total plasma antioxidant capacity. Thus, the putative role of lycopene in combination with other tomato bioactive ingredients in the prevention of oxidative stress related diseases was evidenced.

## Tomato Suppresses the NF-κB Activation and Reduces Inflammation

Inflammation is a normal protective response of the innate immune system to an injury. However, when the oxidative damage is out of control, inflammation may become chronic leading to tissue damage. During inflammation, immune cells are activated and release increased levels of ROS to eliminate invading pathogens. The intracellular ROS production is linked to various cellular processes controlled by NF-κB (which is the central orchestrator of the inflammatory response), including activation of NAD(P)H oxidase, matrix metalloproteinases (MMP-1, MMP-2 and MMP-9), nitric oxide synthases (NOS), xanthine oxidase, cyclooxygenase-2 (COX-2), and lipooxygenases. Additionally, immune cells release a number of proinflammatory mediators such as cytokines (tumour necrosis factors (TNF-α) and interleukin (IL-6, IL-1β), chemokines (IL-8), cell adhesion molecules (CAMs), C-type lectin receptors, prostaglandins, leukotrienes, and NO [110, 111].

Clinical trials demonstrated that the anti-inflammatory effects of tomato compounds are attributed to their capacity to suppress of the activation of NF-κB. De Stefano *et al.* [112] investigated the effect of PS (a polysaccharide from unripe tomato peels) on nitrite and ROS production in J774 macrophages stimulated by bacterial lipopolysaccharide (LPS) for one day. Results demonstrated that PS inhibits NF-κB activation and inducible nitric oxide synthase (iNOS) gene expression by preventing the reactive species production; thus, the involvement of this compound in the control of the oxidative stress and/or inflammation was suggested. Joo *et al.* [113] studied the effects of a tomato lycopene extract on the LPS-induced innate signalling and on the acute and spontaneous chronic intestinal inflammation. Mice were fed a diet containing 0.5 and 2% tomato lycopene extract or an isoflavone free control. The tomato lycopene extract prevented LPS-induced proinflammatory gene expression by blocking of NF-κB signalling, through aggravation of dextran sulfate sodium-induced colitis by enhancing epithelial cell apoptosis. The effectiveness of the combination of carotenoids and phenolics in inhibiting the release of inflammatory mediators from macrophages exposed to LPS and the anti-inflammatory effect in an *in vivo* mouse model of peritonitis was evaluated by Hadad & Levy [7]. Pre-incubation of macrophages with the evaluated compounds, 1 h before the addition of LPS for one day, caused a synergistic inhibition of NO, prostaglandin E(2), and superoxide anion production derived from down-regulation of iNOS, COX-2, and NADPH oxidase protein and mRNA expression and synergistic inhibition of TNF-α secretion. The supplementation of mouse resulted in attenuated neutrophil recruitment to the peritoneal cavity and inhibited inflammatory mediator production by peritoneal neutrophils and macrophages.

**Tomato Reduces Inflammation Linked to Obesity, Diabetes and Cholesterol**

Obesity is a chronic inflammatory state in which the augmented level of body fat leads to an increase in circulating inflammatory mediators [101]. Ghavipour *et al.* [101] demonstrated that tomato juice reduces inflammation in overweight and obese females. In this study, inflammatory biomarkers were analyzed in an intervention group that consumed 330 mL/day of tomato juice for 20 days. Serum levels of IL-8 and TNF-α decreased considerably in the intervention group compared with the control one. Curiously, this effect was confined to overweight subjects. Among obese subjects, the levels of serum IL-6 were reduced in the intervention group, while the levels of IL-8 and TNF-α showed no difference from the control group. Thus, the authors concluded that increased tomato intake may reduce the risk of inflammatory diseases associated with obesity such as cardiovascular disease and diabetes. Another study showed beneficial effects of tomato juice consumption on oxidative stress status of overweight females [11]. Some evidence suggests that oxidative stress, in addition to being a consequence of fat accumulation with subsequent inflammatory response, may be a prerequisite for adipogenesis [114]. The authors verified that the plasma total antioxidant capacity and erythrocyte antioxidant enzymes increased and serum MDA decreased after the 20 days of intervention period. Thus, it was concluded that the verified reduction of oxidative stress in weight females may prevent from obesity related diseases and promote health.

The supplementation effect of tomato juice on indices associated with metabolic health and adipokine profiles in young healthy women, to which was given 280 mL of tomato juice (containing 32.5 mg of lycopene) daily for 2 months, was studied by Li *et al.* [10]. It was found that the tomato juice supplementation significantly reduced the body weight, body fat, body mass index, waist circumference, and the serum levels of cholesterol and TBARS; while the serum levels of adiponectin, triglyceride, and lycopene were significantly increased. Other authors evaluated the effect of pre-meal tomato intake in the anthropometric indices and blood levels of triglycerides, cholesterol, glucose, and uric acid in an intervention group consisting of young adult women [100]. The intervention group ingested a raw ripe tomato (~90 g) before lunch for 4 weeks. At the end of that period, it was observed a positive effect in body weight, fat percentage, and blood levels of glucose, triglycerides, cholesterol, and uric acid of the participants.

Regarding studies in animal models, Seo *et al.* [115] investigated the anti-obesity properties of a tomato vinegar beverage in diet induced obese C57BL/6 mice. The prepared beverage not only reduced fat accumulation, but also insulin resistance; these changes were mediated by the AMP-activated protein kinase and peroxisome proliferator-activated receptor alpha up-regulation. In a similar study conducted by Choi *et al.* [99] it was concluded that green tomato extracts attenuate high-fat diet-induced obesity in C57BL/6 mice through activation of the adenosine monophosphate-activated protein kinase (AMPK) pathway, and that green tomato extracts may be a potential candidate for anti-obesity drugs. Besides, the results indicated that tomatine may be responsible for the observed reduction of body weight. It has been reported that the glycoalkaloid tomatine forms insoluble complexes with cholesterol *in vitro*. In this sense, to evaluate the capacity of tomatine in reducing dietary cholesterol absorption and the plasma levels of cholesterol and triglycerides, hamsters were fed a high-fat, high-cholesterol diet containing 0.05-0.2% of tomatine [116]. The tomatine diet decreased the serum LDL levels without changing HDL cholesterol, being more cholesterol and coprostanol excreted in feces corresponding to the ingested quantity of tomatine. Moreover, these findings suggest that due to the formation and excretion of tomatine-cholesterol complexes, just a very small amount of dietary tomatine is absorbed by the human body.

## Tomato Prevents Cardiovascular Diseases, Atherosclerosis and Hypertension

Currently, cardiovascular disease still represents a major cause of morbidity and death in the world [117]. The endothelium plays a crucial role in cardiovascular health; its dysfunction is associated with the development of atherosclerosis, hypertension and heart failure. Endothelial cells respond to different stimuli through the synthesis and release of several molecules capable of regulating vascular tone, permeability, and inflammation, as well as the blood fluidity and coagulation [118]. Indeed, endothelial dysfunction is a key early step in the development of cardiovascular diseases. It is characterized by an impairment of endothelium-dependent relaxation and a predisposition to a proinflammatory and prothrombotic state [119, 120]. NO contributes to the maintenance of vascular integrity thanks to its antithrombotic, antiatherogenic, and antiproliferative properties. For this reason, decreased levels of NO have been linked with various cardiovascular disorders including hypertension, atherosclerosis and ischaemic disease. The tomato bioactive compounds could contribute to cardiovascular health, and its benefits have been reported in several *in vitro* and *in vivo* studies.

Regarding *in vivo* studies, Kim *et al.* [121] demonstrated that elevated levels of serum lycopene reduce the oxidative stress correlated to endothelial function. Clinical trials were conducted in healthy men who received 15 mg/day of lycopene for 8 weeks. After treatment, serum lycopene levels increased in a dose-dependent manner. The group who received lycopene showed a greater increase in plasma SOD activity and reduction in lymphocyte DNA comet tail length, as well as an increase in reactive hyperemia peripheral arterial tonometry (RH-PAT) index. Moreover, high-sensitivity CRP (hs-CRP), systolic blood pressure, soluble intercellular adhesion molecule-1 (sICAM-1) and soluble vascular cell adhesion molecule-1 (sVCAM-1) were significantly decreased, and β-carotene and LDL-particle size were increased. A remarkable beneficial effect of lycopene supplementation on the endothelial function (*i.e.*, RH-PAT and sVCAM-1) in subjects with an initially relatively impaired endothelial cell function was also observed. Another study showed that the vascular endothelial function of ultra marathon runners is improved by its supplementation with tomato juice for a period of 2 months [8].

The suitability of tomato paste (a concentrated of bioactive compounds) for modifying postprandial oxidative stress, inflammation, and endothelial function in healthy weight individuals was evaluated by Burton-Freeman *et al.* [122]. Participants consumed high-fat meals on two separate occasions containing a processed tomato product or a placebo. Both meals increased the plasma levels of glucose, insulin and lipids. Tomato significantly attenuated the high-fat meal-induced LDL oxidation and the increase in interleukin-6 (IL-6), a proin-flammatory cytokine and inflammation marker. Thus, it was demonstrated that the inclusion of tomatoes or tomato-based foods at the meal reduces the postprandial lipemia-induced oxidative stress and the associated inflammatory response. Furthermore, these findings highlighted the potential protective role of tomato fruit in reducing the risk of cardiovascular disease. Beneficial effects of daily tomato paste consumption on endothelial function were also demonstrated by Xaplanteris *et al.* [123] in a group of healthy young men and women. The tomato supplementation led to an overall flow-mediated dilatation increase particularly in individuals with lower baseline values, and a decrease of total oxidative status past 15 days.

Recently, Vilahur *et al.* [12] reported that the intake of cooked tomato sauce protects against low-density lipoprotein-induced coronary endothelial dysfunction by reducing oxidative damage (diminished DNA damage in the coronary arteries), enhancing endothelial NOS (eNOS) expression and activity, and improving HDL functionality (associated with protein profile changes in apolipoprotein A-I (Apo A-I) and apolipoprotein J (Apo J)). The study was performed in pigs that received a hypercholesterolemic diet and a supplement of cooked tomato sauce (100 g, 21.5 mg lycopene) for 10 days.

Regarding *in vitro* studies, Armoza *et al.* [98] demonstrated the protective role of lycopene and lutein in improving the basic endothelial function. The study was performed in two cultured endothelial cell models and it was verified an increase in the NO levels and a decrease in the endothelin (ET-1) release. Both carotenoids were efficient in attenuating the inflammatory NF-κB signalling, in particular decreasing the TNF-α-induced leukocytes adhesion, the expression of intercellular adhesion molecule-1 (ICAM-1) and vascular cell adhesion molecule-1 (VCAM-1), and nuclear translocation of NF-κB components and some revert of the inhibitor of kappa B (IκB) ubiquitination. Furthermore, both compounds inhibited the NF-κB activation in transfected endothelial cells. This study demonstrated that the prevention of the overexpression of adhesion molecules through inhibition of NF-κB signalling may be one of the main mechanisms driving carotenoids to attenuate inflammatory leukocyte adhesion to endothelium.

## Tomato has Antitumour and Anticarcinogenic Properties

Cancer is a very complex disease caused by cells without the usual control over growth. The apparent causes of this disease can diverge case by case; there are two classes of genes capable of controlling its development, namely oncogenes and tumour suppressor genes. Healthy cells follow a normal growth and proliferation pathway with a definite lifespan, whereas cells with an oncogenic activation undergo much faster division and have an indefinite lifespan. Generally, the cancer formation occurs when an oncogene and a tumour suppressor gene are activated and inactivated in a cell at the same time, respectively. The tumour suppressor genes are involved in the unregulated cell growth inhibition and caretaker genes control the rate of mutation in the genome.

Thus, accumulation of mutation in the genome and consequent higher rate of tumour formation can be caused by a defective caretaker gene. Therefore, cancer arises due to functional deformities in multiple genes [107, 124]. Furthermore, the DNA damage by reactive species can also lead to increased risk of cancer.

In turn, degradation and penetration of the cell extracellular matrix by tumour cells under the action of matrix metalloproteinases (MMPs) and urokinase-type plasminogen activator (u-PA) are key steps in the metastatic cascade of cancer cells. MMPs are the most important proteases involved in tumour cell migration, spreading, tissue invasion and metastasis, and its expression is regulated by extracellular signal regulating kinase (ERK1/2) and c-Jun N-terminal kinase (JNK) [125]. Among them, MMPs, MMP-2 and MMP-9 are key enzymes and are involved in metastasis processes [126]. Furthermore, the inhibition of the mitogen-activated protein kinase (MAPK) pathway (which is involved in signalling cascades) can prevent angiogenesis, proliferation, invasion and metastasis in many tumours [127, 128]. The metastasis process is also regulated by the phosphatidylinositide-3 kinase (PI3K)/Akt signalling pathway [129]. The NF-κB can also facilitate cell proliferation, angiogenesis and metastasis. This protein complex consists of a p50/p65 heterodimer that is masked by the inhibitor I kappa B alpha (IκBα), which causes its retention in the cytoplasm under resting conditions. The IκBα kinase can be activated by various stimuli, including those induced by TNF-α and LPS [95].

Tomato extracts and derived compounds have shown promising effects over different cancer cell lines. Stajčić *et al.* [9] investigated the cell growth activity of tomato waste extracts obtained from different tomato genotypes. Antipro-liferative effects (determined by sulforhodamine B test) were observed in all cell lines (HeLa, MCF7 and MRC-5) at higher concentrations. The authors also correlated the carotenoids content with the antiproliferative and antioxidant activities.

Tang *et al.* [130] verified that low concentration of lycopene and eicosapentaenoic acid could inhibit in a synergistic way the proliferation of human colon cancer HT-29 cells. The inhibitory effects were, partly, associated with the down-regulation of the PI-3K/Akt/mTOR signalling pathway, known to play an important role in tumour progression. Therefore, the inhibition of this pathway is a promising approach for discovery of novel chemotherapeutic agents. More recently, Kim *et al.* [96] investigated the effect of α-tomatine on CT-26 colon cancer cells *in vitro* and *in vivo* in an intracutaneously transplanted mouse tumour. It was demonstrated that α-tomatine in pure form and in tomatine-rich green tomatoes might prevent colon cancer; α-tomatine induced about 50% lysis of the colon cancer cells at 3.5 μM after 24 h of treatment. It was also found that α-tomatine induced cell death through caspases-independent signalling pathways. Intraperitoneally administered α-tomatine also clearly inhibited tumour growth.

Prostate cancer is the second most common cause of male cancer death in the world [131]. A role for NF-κB in the progression of this cancer has been suggested; NF-κB is activated in androgen-insensitive prostate carcinoma cells, and overexpression of NF-κB p65 protein has been detected in the nuclear fraction of prostate cancer clinical specimens [132]. The chemopreventive potential of α-tomatine on androgen-independent human prostatic adenocarcinoma PC-3 cells was evaluated by Lee *et al.* [94]. The treatment with α-tomatine caused a concentration-dependent inhibition of cell growth. It was less cytotoxic to normal human liver WRL-68 cells and normal human prostate RWPE-1 cells. α-Tomatine exhibited its cytotoxic effects against adenocarcinoma PC-3 cells as early as one hour after treatment, which were assessed by the real-time growth kinetics. The glycoalkaloid α-tomatine induced apoptosis and inhibited NF-κB activation, as well as the activation of caspase-3, -8 and -9, suggesting the involvement of both intrinsic and extrinsic apoptosis pathways. Subsequently [95] it was shown that α-tomatine suppresses NF-κB activation through inhibition of IκBα kinase activity, which leads to sequential suppression of IκBα phosphorylation, IκBα degradation, NF-κB p65 phosphorylation, and NF-κB p50/p65 nuclear translocation. As indicated, α-tomatine was able to induce apoptosis; it reduced the TNF-α induced activation of the pro-survival mediator Akt, and the NF-κB inhibition caused a reduction in expression of NF-κB-dependent anti-apoptotic proteins. Moreover, intraperitoneal administration of this bioactive glycoalkaloid clearly attenuated the growth of PC-3 cell tumours (grown subcutaneously and orthotopically) in mice. These effects were accompanied by increased apoptosis, lower proliferation of tumour cells, and low nuclear translocation of the p50 and p65 components of

NF-κB. Recently, Kolberg *et al.* [82] investigated whether tomato paste has the ability to modulate NF-κB activity and cancer-related gene expression in human prostate cancer cells (PC-3) and PC-3 xenografts. PC-3 cells were stably transduced with an NF-κB-luciferase construct and then treated with tomato extract or a placebo. Mice bearing PC-3 xenografts received a high-fat diet with or without 10% tomato paste for 6.5 weeks. The tomato extract considerably inhibited the TNF-α stimulated NF-κB activity in the PC-3 cells, and modulated the expression of genes associated with inflammation, apoptosis, and cancer progression. Mice fed tomato paste diet revealed accumulation of lycopene in liver, xenografts and serum. The tomato paste had no effect on tumour size in mice; but there was a trend toward inhibition of NF-κB activity in the xenografts. Gene expression, most prominent in xenografts, was higher after tomato treatment.

Lung cancer is becoming increasingly common. About 40% of these cancers are adenocarcinoma, a type of non-small cell lung cancers with a low prognosis and highly potential for metastatic [133]. A study carried out by Yan *et al.* [126] examined the effect of tomatidine on the migration and invasion of human lung adenocarcinoma A549 cells. *In vitro* treatments with non-toxic doses of tomatidine resulted in markedly suppressed cell invasion, while cell migration was not affected. Tomatidine reduced the mRNA levels of MMP-2 and MMP-9 and increased the expression of reversion-inducing cysteine-rich protein with kazal motifs (RECK, a protein involved in the proteolytic degradation of extracellular matrix in tumour metastasis), as well as the tissue inhibitor of MMP-1. It also inhibited the ERK and Akt signalling pathways and NF-κB activity.

**Tomato Protects Liver from Hepatotoxicity and Hepatocarcinogenesis**

Oxidative damage caused by free radicals formed during the metabolism of nitrosamines has been suggested as one of the main cause for the initiation of hepatocarcinogenesis [134]. *N*-Nitrosamines are a class of chemical compounds that are metabolised to prooxidant and carcinogenic substances. *N*-Nitrosodieth--lamine is a representative of this class capable of generating carbocations and ROS during the cytochrome P450-mediated biotransformation. Liver injury induced by *N*-nitrosodiethylamine is a well-known model of hepatotoxicity commonly used for screening of hepatoprotective effects of natural matrices. A lycopene-enriched tomato paste tested by Kujawska *et al.* [103] was suitable for suppressing the *N*-nitrosodiethylamine-induced oxidative stress in rats. Pre-treatment with tomato paste protected antioxidant enzymes (SOD, CAT and glutathione reductase) and decreased the DNA damage in leucocytes and the plasma concentration of protein carbonyls. The microsomal lipid peroxidation was also decreased in liver of rats pre-treated with a lower dose of tomato paste. Gupta *et al.* [134] investigated the involvement of tomato lycopene against oxidative stress induced by the deleterious effect of *N*-nitrosodiethylamine on cellular macromolecules of mice, having been demonstrated the intervention of lycopene on the initiation of carcinogenesis. Indeed, lycopene has influence on multiple dysregulated pathways during initiation of carcinogenesis; in particular it helps in the membrane fluidity normalization, improvement of antioxidant enzymes activity and reduction of glutathione (GSH) which accounted for reduced oxidative damage.

The antioxidant and hepatoprotective properties of naringenin and its β-cyclodextrin formulation at a dose of 50 mg per kg of body weight were evaluated by Hermenean *et al.* [102]. Serum-enzymatic and liver antioxidant activity and histopathological and ultrastructural changes were investigated in mice subjected to acute intoxication with carbon tetrachloride ($CCl_4$), one of the most potent hepatotoxins. The authors verified that both naringenin and naringenin/β-cyclodextrin complex have antioxidant and hepatoprotective effects against injuries caused by $CCl_4$. Particularly, 24 h after the $CCl_4$ administration, the activity of the transaminases aspartate aminotransferase (AST) and alanine aminotransferase (ALT) and the levels of MDA were increased. A considerable decrease in SOD, CAT and GPx activities and in the GSH levels were also detected. Additionally, extended centrilobular necrosis, steatosis, fibrosis, and an altered ultrastructure of hepatocytes were also verified.

The primary liver cancer has become the fifth most common malignancy in the world [135]. Due to lack of early detection or screening biomarkers, its diagnosis is made at an advanced stage of the disease. Thus, the identification of potential risk factors for early hepatocarcinogenesis and the search for preventive and/or protective measures against them at an early stage are urgently needed. Growing evidence has associated hepatocellular carcinoma and nonalcoholic steatohepatitis (NASH), a chronic and often "silent" liver disease characterized by fat accumulation and infiltration of inflammatory cells in the liver [136]. Wang *et al.* [137] studied the efficacy of an equivalent dosage of dietary lycopene from either a pure compound or a tomato extract against NASH-induced hepatocarcino-genesis. In this study, rats were injected with diethylnitrosamine and then fed either a Lieber-DeCarli control diet or a high-fat diet with or without lycopene or tomato extract for 6 weeks. Both lycopene and tomato extract supplementations considerably decreased the number of altered hepatic foci, being expressed the placental form of glutathione-S transferase in the liver of rats that received a high-fat diet. Decreased activation of NF-κB was verified. Both supplementations reduced the lipid peroxidation induced by the high-fat diet in the liver; it was observed also a significantly decreased inflammatory foci and mRNA expression of proinflammatory cytokines (TNF-α, IL-1β and IL-12) in the group that received a high-fat diet and the tomato extract. Thus, it was concluded that lycopene and tomato extract inhibit the NASH-induced hepatocarcinogenesis mainly as a result of reduced oxidative stress.

Given the beneficial effect of antioxidant supplementation in metal-induced toxicity, and concerns regarding the benefits of tomatoes on different target tissues, Nwokocha *et al.* [138] elucidated the effect of tomato extract on reducing the accumulation of heavy metal in the liver of rats. The tomato extract administration was beneficial in reducing heavy metal accumulation in the liver, namely reducing uptake and enhancing the elimination of these metals in a time dependent manner. The hepatoprotective effect against cadmium toxicity was very high. Among the tomato bioactive compounds, vitamin C has been reported to decrease liver damage from cadmium, mercury and lead [139, 140].

## SAFETY PRECAUTIONS

*Solanum* is probably the most economically important genus, containing familiar crop species, as well as many species containing poisonous or medicinally useful secondary compounds. Because of this, the tomato plant was long used only for ornamental purposes [1]. However, there are no reports of any toxic effect caused by the consumption of tomato fruit, but some reports correlate potato glycoalkaloids with noxious effects in humans. Potatoes with high levels of glycoalkaloids can severely affect the consumer health or even cause death. Mild poisoning cases can cause headache, vomiting and diarrhea. Some neurological symptoms have also been described, namely drowsiness, visual disturbances, apathy, hallucinations, mental confusion, dizziness, trembling and restlessness [89]. Thus, to ensure safety to consumers, it is necessary to perform further studies of toxicity and bioavailability of the tomato glycoalkaloids considering different maturation stages. Besides glycoalkaloids, other antinutritional and potentially harmful compounds, such as oxalic acid [141] and nitrate [142], have been detected in different concentrations in tomato samples.

## INDUSTRIAL APPLICATIONS

Despite the undeniable importance of tomato in the food processing industry, this fruit as well as its by-products can have other applications on biotechnology, chemical, pharmaceutical, and cosmetic industries. Industrial tomato by-products contain significant amounts of the bioactive compounds endowed with different bioactivities and important health promoting effects, as reported in this chapter [143]. Therefore, as tomato wastes are bioorganic materials and being in line with the trend for sustainability, these value-added molecules can be isolated to be used as natural bioactive ingredients for different industries. Extracts or isolated compounds from tomato by-products can also be used as anti-inflammatory, cardioprotective, anticholesterolemic, antidiabetic, and antitumour agents to develop new products and drugs. The isolated compounds can also be applied in the food industry to develop new functional foods and nutraceuticals, and used as food additives to extend their shelf-life.

## CONCLUDING REMARKS

Tomato is the second most consumed vegetable worldwide, thus being an important source of nourishment for the world's population. This fruit is a dietary source of carotenoids (mainly lycopene), vitamins and phenolic compounds (hydrocinnamic acids and flavonoids) which contribute to its health-promoting effects. Unripe (green) tomatoes also contain glycoalkaloids (α-tomatine), compounds endowed with important bioactive properties. Several *in vitro* and *in vivo* studies carried out in the last years confirm the assigned health-promoting effects to tomato, whether as a processed food, extract or it isolated compounds. These studies highlight its antioxidant and anti-inflammatory effects which, mainly *via* down-regulation of NF-κB and proinflammatory mediators such as cytokines, protect from cardiovascular diseases (avoiding the endothelial dysfunction) and various types of cancer (through the modulation of oncogenic signalling pathways). At the same time, epidemiological studies support the health benefits of the consumption of tomatoes and tomato-based foods. Therefore, this "superfruit" has been considered as a functional and powerful disease-fighting food involved in the prevention of chronic degenerative diseases. However, further studies are necessary to clarify some points related to the bioaccessibility and bioavailability of some bioactive compounds and inherent mechanisms of bioactivity, as well as to evaluate synergistic effects among different compounds in various cell lines and animal models, and optimize extraction conditions of value-added compounds from tomato by-products for industrial applications.

## CONFLICT OF INTEREST

The authors confirm that they have no conflict of interest to declare for this publication.

## ACKNOWLEDGEMENTS

The authors are grateful to the Foundation for Science and Technology (FCT, Portugal) for financial support to CIMO (PEst-OE/AGR/UI0690/2014) and REQUIMTE (PEst-C/EQB/LA0006/2014) and for the scholarship (SFRH/BD /92994/2013) granted to the first author.

## REFERENCES

[1]   Bergougnoux V. The history of tomato: from domestication to biopharming. Biotechnol Adv 2014; 32(1): 170-89.
[http://dx.doi.org/10.1016/j.biotechadv.2013.11.003] [PMID: 24211472]

[2]   Faostat3.fao.org. Food and Agriculture Organization of the United Nations, Statistics Division, 2015. Available from: http://faostat3.fao.org/home/E , [Cited: 4 May 2015];

[3]   Guil-Guerrero JL, Rebolloso-Fuentes MM. Nutrient composition and antioxidant activity of eight tomato (*Lycopersicon esculentum*) varieties. J Food Compos Anal 2009; 22: 123-9.
[http://dx.doi.org/10.1016/j.jfca.2008.10.012]

[4]     Pinela J, Barros L, Carvalho AM, Ferreira IC. Nutritional composition and antioxidant activity of four tomato (*Lycopersicon esculentum* L.) farmer' varieties in Northeastern Portugal homegardens. Food Chem Toxicol 2012; 50(3-4): 829-34.
        [http://dx.doi.org/10.1016/j.fct.2011.11.045] [PMID: 22154854]

[5]     Barros L, Dueñas M, Pinela J, Carvalho AM, Buelga CS, Ferreira IC. Characterization and quantification of phenolic compounds in four tomato (*Lycopersicon esculentum* L.) farmers' varieties in northeastern Portugal homegardens. Plant Foods Hum Nutr 2012; 67(3): 229-34.
        [http://dx.doi.org/10.1007/s11130-012-0307-z] [PMID: 22922837]

[6]     Vallverdú-Queralt A, de Alvarenga JF, Estruch R, Lamuela-Raventos RM. Bioactive compounds present in the Mediterranean sofrito. Food Chem 2013; 141(4): 3365-72.
        [http://dx.doi.org/10.1016/j.foodchem.2013.06.032] [PMID: 23993494]

[7]     Hadad N, Levy R. The synergistic anti-inflammatory effects of lycopene, lutein, β-carotene, and carnosic acid combinations via redox-based inhibition of NF-κB signaling. Free Radic Biol Med 2012; 53(7): 1381-91.
        [http://dx.doi.org/10.1016/j.freeradbiomed.2012.07.078] [PMID: 22889596]

[8]     Samaras A, Tsarouhas K, Paschalidis E, *et al.* Effect of a special carbohydrate-protein bar and tomato juice supplementation on oxidative stress markers and vascular endothelial dynamics in ultra-marathon runners. Food Chem Toxicol 2014; 69: 231-6.
        [http://dx.doi.org/10.1016/j.fct.2014.03.029] [PMID: 24705018]

[9]     Stajčić S, Ćetković G, Čanadanović-Brunet J, Djilas S, Mandić A, Četojević-Simin D. Tomato waste: Carotenoids content, antioxidant and cell growth activities. Food Chem 2015; 172: 225-32.
        [http://dx.doi.org/10.1016/j.foodchem.2014.09.069] [PMID: 25442547]

[10]    Li Y-F, Chang Y-Y, Huang H-C, Wu Y-C, Yang M-D, Chao P-M. Tomato juice supplementation in young women reduces inflammatory adipokine levels independently of body fat reduction. Nutrition 2015; 31(5): 691-6.
        [http://dx.doi.org/10.1016/j.nut.2014.11.008] [PMID: 25837214]

[11]    Ghavipour M, Sotoudeh G, Ghorbani M. Tomato juice consumption improves blood antioxidative biomarkers in overweight and obese females. Clin Nutr 2015; 34(5): 805-9.
        [http://dx.doi.org/10.1016/j.clnu.2014.10.012] [PMID: 25466953]

[12]    Vilahur G, Cubedo J, Padró T, *et al.* Intake of cooked tomato sauce preserves coronary endothelial function and improves apolipoprotein A-I and apolipoprotein J protein profile in high-density lipoproteins. Transl Res 2015; 166(1): 44-56.
        [http://dx.doi.org/10.1016/j.trsl.2014.11.004] [PMID: 25514506]

[13]    García-Valverde V, Navarro-González I, García-Alonso J, Periago MJ. Antioxidant bioactive compounds in selected industrial processing and fresh consumption tomato cultivars. Food Bioprocess Technol 2013; 6(2): 391-402.
        [http://dx.doi.org/10.1007/s11947-011-0687-3]

[14]    Friedman M. Anticarcinogenic, cardioprotective, and other health benefits of tomato compounds lycopene, α-tomatine, and tomatidine in pure form and in fresh and processed tomatoes. J Agric Food Chem 2013; 61(40): 9534-50.
        [http://dx.doi.org/10.1021/jf402654e] [PMID: 24079774]

[15]   Friedman M. Chemistry and anticarcinogenic mechanisms of glycoalkaloids produced by eggplants, potatoes, and tomatoes. J Agric Food Chem 2015; 63(13): 3323-37.
[http://dx.doi.org/10.1021/acs.jafc.5b00818] [PMID: 25821990]

[16]   Aherne SA, Jiwan MA, Daly T, O'Brien NM. Geographical location has greater impact on carotenoid content and bioaccessibility from tomatoes than variety. Plant Foods Hum Nutr 2009; 64(4): 250-6.
[http://dx.doi.org/10.1007/s11130-009-0136-x] [PMID: 19757067]

[17]   Raiola A, Rigano MM, Calafiore R, Frusciante L, Barone A. Enhancing the health-promoting effects of tomato fruit for biofortified food. Mediators Inflamma 2014; 2014: 1-16.
[http://dx.doi.org/10.1155/2014/139873]

[18]   Koh E, Kaffka S, Mitchell AE. A long-term comparison of the influence of organic and conventional crop management practices on the content of the glycoalkaloid α-tomatine in tomatoes. J Sci Food Agric 2013; 93(7): 1537-42.
[http://dx.doi.org/10.1002/jsfa.5951] [PMID: 23138335]

[19]   Masetti O, Ciampa A, Nisini L, Valentini M, Sequi P, Dell'Abate MT. Cherry tomatoes metabolic profile determined by [1]H-High Resolution-NMR spectroscopy as influenced by growing season. Food Chem 2014; 162: 215-22.
[http://dx.doi.org/10.1016/j.foodchem.2014.04.066] [PMID: 24874378]

[20]   Iglesias MJ, García-López J, Collados-Luján JF, *et al.* Differential response to environmental and nutritional factors of high-quality tomato varieties. Food Chem 2015; 176: 278-87.
[http://dx.doi.org/10.1016/j.foodchem.2014.12.043] [PMID: 25624234]

[21]   Namitha KK, Negi PS. Chemistry and biotechnology of carotenoids. Crit Rev Food Sci Nutr 2010; 50(8): 728-60.
[http://dx.doi.org/10.1080/10408398.2010.499811] [PMID: 20830634]

[22]   Palozza P, Catalano A, Simone RE, Mele MC, Cittadini A. Effect of lycopene and tomato products on cholesterol metabolism. Ann Nutr Metab 2012; 61(2): 126-34.
[http://dx.doi.org/10.1159/000342077] [PMID: 22965217]

[23]   Shi J, Le Maguer M. Lycopene in tomatoes: chemical and physical properties affected by food processing. Crit Rev Biotechnol 2000; 20(4): 293-334.
[http://dx.doi.org/10.1080/07388550091144212] [PMID: 11192026]

[24]   Tohge T, Alseekh S, Fernie AR. On the regulation and function of secondary metabolism during fruit development and ripening. J Exp Bot 2014; 65(16): 4599-611.
[http://dx.doi.org/10.1093/jxb/ert443] [PMID: 24446507]

[25]   Hernandez-Marin E, Galano A, Martínez A. Cis carotenoids: colorful molecules and free radical quenchers. J Phys Chem B 2013; 117(15): 4050-61.
[http://dx.doi.org/10.1021/jp401647n] [PMID: 23560647]

[26]   Zaripheh S, Boileau TW, Lila MA, Erdman JW Jr. [$^{14}$C]-lycopene and [$^{14}$C]-labeled polar products are differentially distributed in tissues of F344 rats prefed lycopene. J Nutr 2003; 133(12): 4189-95.
[PMID: 14652370]

[27]   Navarro-González I, Pérez-Sánchez H, Martín-Pozuelo G, García-Alonso J, Periago MJ. The inhibitory effects of bioactive compounds of tomato juice binding to hepatic HMGCR: *in vivo* study

and molecular modelling. PLoS One 2014; 9(1): e83968.
[http://dx.doi.org/10.1371/journal.pone.0083968] [PMID: 24392102]

[28] Boileau TW, Boileau AC, Erdman JW Jr. Bioavailability of all-trans and cis-isomers of lycopene. Exp Biol Med (Maywood) 2002; 227(10): 914-9.
[PMID: 12424334]

[29] Zhang J, Hou X, Ahmad H, Zhang H, Zhang L, Wang T. Assessment of free radicals scavenging activity of seven natural pigments and protective effects in AAPH-challenged chicken erythrocytes. Food Chem 2014; 145: 57-65.
[http://dx.doi.org/10.1016/j.foodchem.2013.08.025] [PMID: 24128449]

[30] Darrigues A, Schwartz SJ, Francis DM. Optimizing sampling of tomato fruit for carotenoid content with application to assessing the impact of ripening disorders. J Agric Food Chem 2008; 56(2): 483-7.
[http://dx.doi.org/10.1021/jf071896v] [PMID: 18092756]

[31] Viskelis P, Jankauskiene J, Bobinaite R. Content of carotenoids and physical properties of tomatoes harvested at different ripening stages. In: Proceedings of the 3rd Baltic Conference on Food Science and Technology - FOODBALT; April 17-18; Jelgava, Latvia. 2008.

[32] Defo MA, Spear PA, Couture P. Consequences of metal exposure on retinoid metabolism in vertebrates: a review. Toxicol Lett 2014; 225(1): 1-11.
[http://dx.doi.org/10.1016/j.toxlet.2013.11.024] [PMID: 24291063]

[33] Khachik F, Carvalho L, Bernstein PS, Muir GJ, Zhao DY, Katz NB. Chemistry, distribution, and metabolism of tomato carotenoids and their impact on human health. Exp Biol Med (Maywood) 2002; 227(10): 845-51.
[PMID: 12424324]

[34] Baiano A, Del Nobile MA. Antioxidant compounds from vegetable matrices: Biosynthesis, occurrence, and extraction systems. Crit Rev Food Sci Nutr 2015. In press.
[PMID: 25751787]

[35] Davey MW, Van Montagu M, Inzé D, et al. Plant_L-ascorbic acid: chemistry, function, metabolism, bioavailability and effects of processing. J Sci Food Agric 2000; 80(7): 825-60.
[http://dx.doi.org/10.1002/(SICI)1097-0010(20000515)80:7<825::AID-JSFA598>3.0.CO;2-6]

[36] Lee J, Koo N, Min DB. Reactive oxygen species, aging, and antioxidative nutraceuticals. Compr Rev Food Sci F 2004; 3(1): 21-33.
[http://dx.doi.org/10.1111/j.1541-4337.2004.tb00058.x]

[37] Abushita AA, Daood HG, Biacs PA. Change in carotenoids and antioxidant vitamins in tomato as a function of varietal and technological factors. J Agric Food Chem 2000; 48(6): 2075-81.
[http://dx.doi.org/10.1021/jf990715p] [PMID: 10888501]

[38] Frusciante L, Carli P, Ercolano MR, et al. Antioxidant nutritional quality of tomato. Mol Nutr Food Res 2007; 51(5): 609-17.
[http://dx.doi.org/10.1002/mnfr.200600158] [PMID: 17427261]

[39] Carocho M, Ferreira IC. A review on antioxidants, prooxidants and related controversy: natural and synthetic compounds, screening and analysis methodologies and future perspectives. Food Chem Toxicol 2013; 51: 15-25.

[http://dx.doi.org/10.1016/j.fct.2012.09.021] [PMID: 23017782]

[40]    Burton GW, Traber MG. Vitamin E: antioxidant activity, biokinetics, and bioavailability. Annu Rev Nutr 1990; 10: 357-82.
[http://dx.doi.org/10.1146/annurev.nu.10.070190.002041] [PMID: 2200468]

[41]    Packer L, Weber SU, Rimbach G. Molecular aspects of α-tocotrienol antioxidant action and cell signalling. J Nutr 2001; 131(2): 369S-73S.
[PMID: 11160563]

[42]    Giovannoni JJ. Completing a pathway to plant vitamin C synthesis. Proc Natl Acad Sci USA 2007; 104(22): 9109-10.
[http://dx.doi.org/10.1073/pnas.0703222104] [PMID: 17517613]

[43]    Martin H, Comeskey D, Simpson RM, Laing WA, McGhie TK. Quantification of folate in fruits and vegetables: A fluorescence-based homogeneous assay. Anal Biochem 2010; 402(2): 137-45.
[http://dx.doi.org/10.1016/j.ab.2010.03.032] [PMID: 20361923]

[44]    Tyagi K, Upadhyaya P, Sarma S, Tamboli V, Sreelakshmi Y, Sharma R. High performance liquid chromatography coupled to mass spectrometry for profiling and quantitative analysis of folate monoglutamates in tomato. Food Chem 2015; 179: 76-84.
[http://dx.doi.org/10.1016/j.foodchem.2015.01.110] [PMID: 25722141]

[45]    Morales P, Fernández-Ruiz V, Sánchez-Mata MC, Cámara M, Tardío J. Optimization and application of FL-HPLC for folates analysis in 20 species of Mediterranean wild vegetables. Food Anal Methods 2015; 8(2): 302-11.
[http://dx.doi.org/10.1007/s12161-014-9887-6]

[46]    Forges T, Monnier-Barbarino P, Alberto JM, Guéant-Rodriguez RM, Daval JL, Guéant JL. Impact of folate and homocysteine metabolism on human reproductive health. Hum Reprod Update 2007; 13(3): 225-38.
[http://dx.doi.org/10.1093/humupd/dml063] [PMID: 17307774]

[47]    Wagner C. Biochemical role of folate in cellular metabolism. In: Bailey LB, Ed. Folate in health and disease. New York: Marcel Dekker 1995; pp. 23-42.

[48]    Selhub J, Jacques PF, Wilson PW, Rush D, Rosenberg IH. Vitamin status and intake as primary determinants of homocysteinemia in an elderly population. JAMA 1993; 270(22): 2693-8.
[http://dx.doi.org/10.1001/jama.1993.03510220049033] [PMID: 8133587]

[49]    Hackam DG, Anand SS. Emerging risk factors for atherosclerotic vascular disease: a critical review of the evidence. JAMA 2003; 290(7): 932-40.
[http://dx.doi.org/10.1001/jama.290.7.932] [PMID: 12928471]

[50]    Splaver A, Lamas GA, Hennekens CH. Homocysteine and cardiovascular disease: biological mechanisms, observational epidemiology, and the need for randomized trials A Heart J 2004; 148(1): 34-40.

[51]    Escarpa A, Gonzalez MC. An overview of analytical chemistry of phenolic compounds in foods. Crit Rev Anal Chem 2001; 31: 57-139.
[http://dx.doi.org/10.1080/20014091076695]

[52]    Vallverdú-Queralt A, Jáuregui O, Medina-Remón A, Andrés-Lacueva C, Lamuela-Raventós RM.

Improved characterization of tomato polyphenols using liquid chromatography/electrospray ionization linear ion trap quadrupole Orbitrap mass spectrometry and liquid chromatography/electrospray ionization tandem mass spectrometry. Rapid Commun Mass Spectrom 2010; 24(20): 2986-92.
[http://dx.doi.org/10.1002/rcm.4731] [PMID: 20872631]

[53]    Sato Y, Itagaki S, Kurokawa T, *et al. In vitro* and *in vivo* antioxidant properties of chlorogenic acid and caffeic acid. Int J Pharm 2011; 403(1-2): 136-8.
[http://dx.doi.org/10.1016/j.ijpharm.2010.09.035] [PMID: 20933071]

[54]    Kono Y, Shibata H, Kodama Y, Sawa Y. The suppression of the N-nitrosating reaction by chlorogenic acid. Biochem J 1995; 312(Pt 3): 947-53.
[http://dx.doi.org/10.1042/bj3120947] [PMID: 8554543]

[55]    Le Gall G, DuPont MS, Mellon FA, *et al.* Characterization and content of flavonoid glycosides in genetically modified tomato (*Lycopersicon esculentum*) fruits. J Agric Food Chem 2003; 51(9): 2438-46.
[http://dx.doi.org/10.1021/jf025995e] [PMID: 12696918]

[56]    Bahorun T, Luximon-Ramma A, Crozier A, Aruoma OI. Total phenol, flavonoid, proanthocyanidin and vitamin C levels and antioxidant activities of Mauritian vegetables. J Sci Food Agric 2004; 84(12): 1553-61.
[http://dx.doi.org/10.1002/jsfa.1820]

[57]    Slimestad R, Fossen T, Verheul MJ. The flavonoids of tomatoes. J Agric Food Chem 2008; 56(7): 2436-41.
[http://dx.doi.org/10.1021/jf073434n] [PMID: 18318499]

[58]    Gómez-Romero M, Segura-Carretero A, Fernández-Gutiérrez A. Metabolite profiling and quantification of phenolic compounds in methanol extracts of tomato fruit. Phytochemistry 2010; 71(16): 1848-64.
[http://dx.doi.org/10.1016/j.phytochem.2010.08.002] [PMID: 20810136]

[59]    Friedman M. Analysis of biologically active compounds in potatoes (*Solanum tuberosum*), tomatoes (*Lycopersicon esculentum*), and jimson weed (*Datura stramonium*) seeds. J Chromatogr A 2004; 1054(1-2): 143-55.
[http://dx.doi.org/10.1016/j.chroma.2004.04.049] [PMID: 15553139]

[60]    Fujiwara Y, Takaki A, Uehara Y, *et al.* Tomato steroidal alkaloid glycosides, esculeosides A and B, from ripe fruits. Tetrahedron 2004; 60(22): 4915-20.
[http://dx.doi.org/10.1016/j.tet.2004.03.088]

[61]    Nohara T, Ono M, Ikeda T, Fujiwara Y, El-Aasr M. The tomato saponin, esculeoside a. J Nat Prod 2010; 73(10): 1734-41.
[http://dx.doi.org/10.1021/np100311t] [PMID: 20853874]

[62]    Iijima Y, Nakamura Y, Ogata Y, *et al.* Metabolite annotations based on the integration of mass spectral information. Plant J 2008; 54(5): 949-62.
[http://dx.doi.org/10.1111/j.1365-313X.2008.03434.x] [PMID: 18266924]

[63]    Carbonell-Capella JM, Buniowska M, Barba FJ, Esteve MJ, Frígola A. Analytical methods for determining bioavailability and bioaccessibility of bioactive compounds from fruits and vegetables: a review. Comp Rev Food Sci F 2014; 13(2): 155-71.

[http://dx.doi.org/10.1111/1541-4337.12049]

[64]     Rich GT, Faulks RM, Wickham MS, Fillery-Travis A. Solubilization of carotenoids from carrot juice and spinach in lipid phases: II. Modeling the duodenal environment. Lipids 2003; 38(9): 947-56.
[http://dx.doi.org/10.1007/s11745-003-1148-z] [PMID: 14584602]

[65]     Story EN, Kopec RE, Schwartz SJ, Harris GK. An update on the health effects of tomato lycopene. Annu Rev Food Sci Technol 2010; 1: 189-210.
[http://dx.doi.org/10.1146/annurev.food.102308.124120] [PMID: 22129335]

[66]     Fernández-García E, Carvajal-Lérida I, Jarén-Galán M, Garrido-Fernández J, Pérez-Gálvez A, Hornero-Méndez D. Carotenoids bioavailability from foods: from plant pigments to efficient biological activities. Food Res Int 2012; 46(2): 438-50.
[http://dx.doi.org/10.1016/j.foodres.2011.06.007]

[67]     Arranz S, Martínez-Huélamo M, Vallverdu-Queralt A, *et al.* Influence of olive oil on carotenoid absorption from tomato juice and effects on postprandial lipemia. Food Chem 2015; 168: 203-10.
[http://dx.doi.org/10.1016/j.foodchem.2014.07.053] [PMID: 25172701]

[68]     Unlu NZ, Bohn T, Francis D, Clinton SK, Schwartz SJ. Carotenoid absorption in humans consuming tomato sauces obtained from tangerine or high-beta-carotene varieties of tomatoes. J Agric Food Chem 2007; 55(4): 1597-603.
[http://dx.doi.org/10.1021/jf062337b] [PMID: 17243700]

[69]     Lodge JK. Vitamin E bioavailability in humans. J Plant Physiol 2005; 162(7): 790-6.
[http://dx.doi.org/10.1016/j.jplph.2005.04.012] [PMID: 16008106]

[70]     Martin KR, Appel CL. Polyphenols as dietary supplements: a double-edged sword. Nutr Diet Suppl 2010; 2: 1-12.

[71]     Heleno SA, Martins A, Queiroz MJ, Ferreira IC. Bioactivity of phenolic acids: metabolites *versus* parent compounds: a review. Food Chem 2015; 173: 501-13.
[http://dx.doi.org/10.1016/j.foodchem.2014.10.057] [PMID: 25466052]

[72]     Bugianesi R, Salucci M, Leonardi C, *et al.* Effect of domestic cooking on human bioavailability of naringenin, chlorogenic acid, lycopene and β-carotene in cherry tomatoes. Eur J Nutr 2004; 43(6): 360-6.
[http://dx.doi.org/10.1007/s00394-004-0483-1] [PMID: 15309458]

[73]     Martínez-Huélamo M, Tulipani S, Estruch R, *et al.* The tomato sauce making process affects the bioaccessibility and bioavailability of tomato phenolics: a pharmacokinetic study. Food Chem 2015; 173: 864-72.
[http://dx.doi.org/10.1016/j.foodchem.2014.09.156] [PMID: 25466100]

[74]     Hoffmann D, Fnimh AH, Eds. Medical herbalism: the science and practice of herbal medicine. Vermont: Healing Arts Press 2003.

[75]     Pham-Huy LA, He H, Pham-Huy C. Free radicals, antioxidants in disease and health. Int J Biomed Sci 2008; 4(2): 89-96.
[PMID: 23675073]

[76]     Hayden MS, West AP, Ghosh S. NF-kappaB and the immune response. Oncogene 2006; 25(51): 6758-80.

[http://dx.doi.org/10.1038/sj.onc.1209943] [PMID: 17072327]

[77]   Nimse SN, Pal D. Free radicals, natural antioxidants, and their reaction mechanisms. RSC Advances 2015; 5(35): 27986-8006.
[http://dx.doi.org/10.1039/C4RA13315C]

[78]   Paiva SA, Russell RM. β-carotene and other carotenoids as antioxidants. J Am Coll Nutr 1999; 18(5): 426-33.
[http://dx.doi.org/10.1080/07315724.1999.10718880] [PMID: 10511324]

[79]   Heber D, Lu QY. Overview of mechanisms of action of lycopene. Exp Biol Med (Maywood) 2002; 227(10): 920-3.
[PMID: 12424335]

[80]   Kim G-Y, Kim J-H, Ahn S-C, *et al.* Lycopene suppresses the lipopolysaccharide-induced phenotypic and functional maturation of murine dendritic cells through inhibition of mitogen-activated protein kinases and nuclear factor-kappaB. Immunology 2004; 113(2): 203-11.
[http://dx.doi.org/10.1111/j.1365-2567.2004.01945.x] [PMID: 15379981]

[81]   Reifen R, Nissenkorn A, Matas Z, Bujanover Y. 5-ASA and lycopene decrease the oxidative stress and inflammation induced by iron in rats with colitis. J Gastroenterol 2004; 39(6): 514-9.
[http://dx.doi.org/10.1007/s00535-003-1336-z] [PMID: 15235867]

[82]   Kolberg M, Pedersen S, Bastani NE, Carlsen H, Blomhoff R, Paur I. Tomato paste alters NF-κB and cancer-related mRNA expression in prostate cancer cells, xenografts, and xenograft microenvironment. Nutr Cancer 2015; 67(2): 305-15.
[http://dx.doi.org/10.1080/01635581.2015.990575] [PMID: 25664890]

[83]   Morlière P, Patterson LK, Santos CM, *et al.* The dependence of α-tocopheroxyl radical reduction by hydroxy-2,3-diarylxanthones on structure and micro-environment. Org Biomol Chem 2012; 10(10): 2068-76.
[http://dx.doi.org/10.1039/c2ob06612b] [PMID: 22302132]

[84]   Niki E. Role of vitamin E as a lipid-soluble peroxyl radical scavenger: *in vitro* and *in vivo* evidence. Free Radic Biol Med 2014; 66: 3-12.
[http://dx.doi.org/10.1016/j.freeradbiomed.2013.03.022] [PMID: 23557727]

[85]   Livrea MA, Tesoriere L, Bongiorno A, Pintaudi AM, Ciaccio M, Riccio A. Contribution of vitamin A to the oxidation resistance of human low density lipoproteins. Free Radic Biol Med 1995; 18(3): 401-9.
[http://dx.doi.org/10.1016/0891-5849(94)00151-9] [PMID: 9101230]

[86]   Meyer AS, Donovan JL, Pearson DA, Waterhouse AL, Frankel EN. Fruit hydroxycinnamic acids inhibit human low-density lipoprotein oxidation *in vitro*. J Agric Food Chem 1998; 46(5): 1783-7.
[http://dx.doi.org/10.1021/jf9708960]

[87]   Procházková D, Boušová I, Wilhelmová N. Antioxidant and prooxidant properties of flavonoids. Fitoterapia 2011; 82(4): 513-23.
[http://dx.doi.org/10.1016/j.fitote.2011.01.018] [PMID: 21277359]

[88]   Ihme N, Kiesewetter H, Jung F, *et al.* Leg oedema protection from a buckwheat herb tea in patients with chronic venous insufficiency: a single-centre, randomised, double-blind, placebo-controlled

clinical trial. Eur J Clin Pharmacol 1996; 50(6): 443-7.
[http://dx.doi.org/10.1007/s002280050138] [PMID: 8858269]

[89]  Milner SE, Brunton NP, Jones PW, O'Brien NM, Collins SG, Maguire AR. Bioactivities of glycoalkaloids and their aglycones from *Solanum* species. J Agric Food Chem 2011; 59(8): 3454-84.
[http://dx.doi.org/10.1021/jf200439q] [PMID: 21401040]

[90]  Fletcher SP, Geyer BC, Smith A, *et al.* Tissue distribution of cholinesterases and anticholinesterases in native and transgenic tomato plants. Plant Mol Biol 2004; 55(1): 33-43.
[http://dx.doi.org/10.1007/s11103-004-0394-9] [PMID: 15604663]

[91]  Alozie SO, Sharma RP, Salunkhe DK. Inhibition of rat cholinesterase isoenzymes *in vitro* and *in vivo* by the potato alkaloid, α-chaconine. J Food Biochem 1978; 2(3): 259-76.
[http://dx.doi.org/10.1111/j.1745-4514.1978.tb00621.x]

[92]  Roddick JG, Weissenberg M, Leonard AL. Membrane disruption and enzyme inhibition by naturally-occurring and modified chacotriose-containing *Solanum* steroidal glycoalkaloids. Phytochemistry 2001; 56(6): 603-10.
[http://dx.doi.org/10.1016/S0031-9422(00)00420-9] [PMID: 11281138]

[93]  Chiu FL, Lin JK. Tomatidine inhibits iNOS and COX-2 through suppression of NF-kappaB and JNK pathways in LPS-stimulated mouse macrophages. FEBS Lett 2008; 582(16): 2407-12.
[http://dx.doi.org/10.1016/j.febslet.2008.05.049] [PMID: 18544347]

[94]  Lee S-T, Wong P-F, Cheah S-C, Mustafa MR. Alpha-tomatine induces apoptosis and inhibits nuclear factor-kappa B activation on human prostatic adenocarcinoma PC-3 cells. PLoS One 2011; 6(4): e18915.
[http://dx.doi.org/10.1371/journal.pone.0018915] [PMID: 21541327]

[95]  Lee S-T, Wong P-F, He H, Hooper JD, Mustafa MR. Alpha-tomatine attenuation of *in vivo* growth of subcutaneous and orthotopic xenograft tumors of human prostate carcinoma PC-3 cells is accompanied by inactivation of nuclear factor-kappa B signaling. PLoS One 2013; 8(2): e57708.
[http://dx.doi.org/10.1371/journal.pone.0057708] [PMID: 23437404]

[96]  Kim SP, Nam SH, Friedman M. The tomato glycoalkaloid α-tomatine induces caspase-independent cell death in mouse colon cancer CT-26 cells and transplanted tumors in mice. J Agric Food Chem 2015; 63: 1142-50.
[http://dx.doi.org/10.1021/jf5040288] [PMID: 25614934]

[97]  Kúdelová J, Seifrtová M, Suchá L, Tomšík P, Havelek R, Řezáčová M. Alpha-tomatine activates cell cycle checkpoints in the absence of DNA damage in human leukemic MOLT-4 cells. J Appl Biomed 2013; 11(2): 93-103.
[http://dx.doi.org/10.2478/v10136-012-0033-8]

[98]  Armoza A, Haim Y, Bashiri A, Wolak T, Paran E. Tomato extract and the carotenoids lycopene and lutein improve endothelial function and attenuate inflammatory NF-κB signaling in endothelial cells. J Hypertens 2013; 31(3): 521-9.
[http://dx.doi.org/10.1097/HJH.0b013e32835c1d01] [PMID: 23235359]

[99]  Choi KM, Lee YS, Shin DM, *et al.* Green tomato extract attenuates high-fat-diet-induced obesity through activation of the AMPK pathway in C57BL/6 mice. J Nutr Biochem 2013; 24(1): 335-42.
[http://dx.doi.org/10.1016/j.jnutbio.2012.06.018] [PMID: 22974972]

[100] Vinha AF, Barreira SV, Costa AS, Alves RC, Oliveira MB. Pre-meal tomato (*Lycopersicon esculentum*) intake can have anti-obesity effects in young women? Int J Food Sci Nutr 2014; 65(8): 1019-26.
[http://dx.doi.org/10.3109/09637486.2014.950206] [PMID: 25156566]

[101] Ghavipour M, Saedisomeolia A, Djalali M, *et al.* Tomato juice consumption reduces systemic inflammation in overweight and obese females. Br J Nutr 2013; 109(11): 2031-5.
[http://dx.doi.org/10.1017/S0007114512004278] [PMID: 23069270]

[102] Hermenean A, Ardelean A, Stan M, *et al.* Antioxidant and hepatoprotective effects of naringenin and its β-cyclodextrin formulation in mice intoxicated with carbon tetrachloride: a comparative study. J Med Food 2014; 17(6): 670-7.
[http://dx.doi.org/10.1089/jmf.2013.0007] [PMID: 24611872]

[103] Kujawska M, Ewertowska M, Adamska T, Sadowski C, Ignatowicz E, Jodynis-Liebert J. Antioxidant effect of lycopene-enriched tomato paste on *N*-nitrosodiethylamine-induced oxidative stress in rats. J Physiol Biochem 2014; 70(4): 981-90.
[http://dx.doi.org/10.1007/s13105-014-0367-7] [PMID: 25387411]

[104] Li H, Deng Z, Liu R, Loewen S, Tsao R. Bioaccessibility, *in vitro* antioxidant activities and *in vivo* anti-inflammatory activities of a purple tomato (*Solanum lycopersicum* L.). Food Chem 2014; 159: 353-60.
[http://dx.doi.org/10.1016/j.foodchem.2014.03.023] [PMID: 24767066]

[105] Visioli F, Riso P, Grande S, Galli C, Porrini M. Protective activity of tomato products on *in vivo* markers of lipid oxidation. Eur J Nutr 2003; 42(4): 201-6.
[http://dx.doi.org/10.1007/s00394-003-0415-5] [PMID: 12923651]

[106] Uttara B, Singh AV, Zamboni P, Mahajan RT. Oxidative stress and neurodegenerative diseases: a review of upstream and downstream antioxidant therapeutic options. Curr Neuropharmacol 2009; 7(1): 65-74.
[http://dx.doi.org/10.2174/157015909787602823] [PMID: 19721819]

[107] Pal D, Banerjee S, Ghosh AK. Dietary-induced cancer prevention: An expanding research arena of emerging diet related to healthcare system. J Adv Pharm Technol Res 2012; 3(1): 16-24.
[PMID: 22470889]

[108] Marrazzo G, Barbagallo I, Galvano F, *et al.* Role of dietary and endogenous antioxidants in diabetes. Crit Rev Food Sci Nutr 2014; 54(12): 1599-616.
[http://dx.doi.org/10.1080/10408398.2011.644874] [PMID: 24580561]

[109] Abete I, Perez-Cornago A, Navas-Carretero S, Bondia-Pons I, Zulet MA, Martinez JÁ. A regular lycopene enriched tomato sauce consumption influences antioxidant status of healthy young-subjects: a crossover study. J Funct Foods 2013; 5(1): 28-35.
[http://dx.doi.org/10.1016/j.jff.2012.07.007]

[110] Oka T, Fujimoto M, Nagasaka R, Ushio H, Hori M, Ozaki H. Cycloartenyl ferulate, a component of rice bran oil-derived γ-oryzanol, attenuates mast cell degranulation. Phytomedicine 2010; 17(2): 152-6.
[http://dx.doi.org/10.1016/j.phymed.2009.05.013] [PMID: 19577449]

[111] Shalini V, Bhaskar S, Kumar KS, Mohanlal S, Jayalekshmy A, Helen A. Molecular mechanisms of anti-inflammatory action of the flavonoid, tricin from Njavara rice (Oryza sativa L.) in human peripheral blood mononuclear cells: possible role in the inflammatory signaling. Int Immunopharmacol 2012; 14(1): 32-8.
[http://dx.doi.org/10.1016/j.intimp.2012.06.005] [PMID: 22705359]

[112] De Stefano D, Tommonaro G, Simeon V, Poli A, Nicolaus B, Carnuccio R. A Polysaccharide from tomato (Lycopersicon esculentum) peels affects NF-kappaB activation in LPS-stimulated J774 macrophages. J Nat Prod 2007; 70(10): 1636-9.
[http://dx.doi.org/10.1021/np070168z] [PMID: 17764147]

[113] Joo Y-E, Karrasch T, Mühlbauer M, *et al.* Tomato lycopene extract prevents lipopolysaccharide-induced NF-kappaB signaling but worsens dextran sulfate sodium-induced colitis in NF-kappaBEGFP mice. PLoS One 2009; 4(2): e4562.
[http://dx.doi.org/10.1371/journal.pone.0004562] [PMID: 19234608]

[114] Tormos KV, Anso E, Hamanaka RB, *et al.* Mitochondrial complex III ROS regulate adipocyte differentiation. Cell Metab 2011; 14(4): 537-44.
[http://dx.doi.org/10.1016/j.cmet.2011.08.007] [PMID: 21982713]

[115] Seo KI, Lee J, Choi RY, *et al.* Anti-obesity and anti-insulin resistance effects of tomato vinegar beverage in diet-induced obese mice. Food Funct 2014; 5(7): 1579-86.
[http://dx.doi.org/10.1039/c4fo00135d] [PMID: 24867606]

[116] Friedman M, Fitch TE, Yokoyama WE. Lowering of plasma LDL cholesterol in hamsters by the tomato glycoalkaloid tomatine. Food Chem Toxicol 2000; 38(7): 549-53.
[http://dx.doi.org/10.1016/S0278-6915(00)00050-8] [PMID: 10942315]

[117] McCullough ML, Peterson JJ, Patel R, Jacques PF, Shah R, Dwyer JT. Flavonoid intake and cardiovascular disease mortality in a prospective cohort of US adults. Am J Clin Nutr 2012; 95(2): 454-64.
[http://dx.doi.org/10.3945/ajcn.111.016634] [PMID: 22218162]

[118] Widlansky ME, Gokce N, Keaney JF Jr, Vita JA. The clinical implications of endothelial dysfunction. J Am Coll Cardiol 2003; 42(7): 1149-60.
[http://dx.doi.org/10.1016/S0735-1097(03)00994-X] [PMID: 14522472]

[119] Herrmann J, Lerman A. The endothelium - the cardiovascular health barometer. Herz 2008; 33(5): 343-53.
[http://dx.doi.org/10.1007/s00059-008-3088-2] [PMID: 18773154]

[120] Vanhoutte PM. Endothelial dysfunction: the first step toward coronary arteriosclerosis. Circ J 2009; 73(4): 595-601.
[http://dx.doi.org/10.1253/circj.CJ-08-1169] [PMID: 19225203]

[121] Kim JY, Paik JK, Kim OY, *et al.* Effects of lycopene supplementation on oxidative stress and markers of endothelial function in healthy men. Atherosclerosis 2011; 215(1): 189-95.
[http://dx.doi.org/10.1016/j.atherosclerosis.2010.11.036] [PMID: 21194693]

[122] Burton-Freeman B, Talbot J, Park E, Krishnankutty S, Edirisinghe I. Protective activity of processed tomato products on postprandial oxidation and inflammation: a clinical trial in healthy weight men and

women. Mol Nutr Food Res 2012; 56(4): 622-31.
[http://dx.doi.org/10.1002/mnfr.201100649] [PMID: 22331646]

[123] Xaplanteris P, Vlachopoulos C, Pietri P, *et al.* Tomato paste supplementation improves endothelial dynamics and reduces plasma total oxidative status in healthy subjects. Nutr Res 2012; 32(5): 390-4.
[http://dx.doi.org/10.1016/j.nutres.2012.03.011] [PMID: 22652379]

[124] Weed DL. The quality of nutrition and cancer reviews: a systematic assessment. Crit Rev Food Sci Nutr 2013; 53(3): 276-86.
[http://dx.doi.org/10.1080/10408398.2010.523853] [PMID: 23215999]

[125] Itoh Y, Nagase H. Matrix metalloproteinases in cancer. Essays Biochem 2002; 38: 21-36.
[http://dx.doi.org/10.1042/bse0380021] [PMID: 12463159]

[126] Yan K-H, Lee L-M, Yan S-H, *et al.* Tomatidine inhibits invasion of human lung adenocarcinoma cell A549 by reducing matrix metalloproteinases expression. Chem Biol Interact 2013; 203(3): 580-7.
[http://dx.doi.org/10.1016/j.cbi.2013.03.016] [PMID: 23566884]

[127] Chien CS, Shen KH, Huang JS, Ko SC, Shih YW. Antimetastatic potential of fisetin involves inactivation of the PI3K/Akt and JNK signaling pathways with downregulation of MMP-2/9 expressions in prostate cancer PC-3 cells. Mol Cell Biochem 2010; 333(1-2): 169-80.
[http://dx.doi.org/10.1007/s11010-009-0217-z] [PMID: 19633975]

[128] Chen PS, Shih YW, Huang HC, Cheng HW. Diosgenin, a steroidal saponin, inhibits migration and invasion of human prostate cancer PC-3 cells by reducing matrix metalloproteinases expression. PLoS One 2011; 6(5): e20164.
[http://dx.doi.org/10.1371/journal.pone.0020164] [PMID: 21629786]

[129] Shukla S, Maclennan GT, Hartman DJ, Fu P, Resnick MI, Gupta S. Activation of PI3K-Akt signaling pathway promotes prostate cancer cell invasion. Int J Cancer 2007; 121(7): 1424-32.
[http://dx.doi.org/10.1002/ijc.22862] [PMID: 17551921]

[130] Tang F-Y, Cho H-J, Pai M-H, Chen Y-H. Concomitant supplementation of lycopene and eicosapentaenoic acid inhibits the proliferation of human colon cancer cells. J Nutr Biochem 2009; 20(6): 426-34.
[http://dx.doi.org/10.1016/j.jnutbio.2008.05.001] [PMID: 18708285]

[131] Food, nutrition, physical activity, and the prevention of cancer: a global perspective. Washington, DC: AICR 2007.

[132] Yemelyanov A, Gasparian A, Lindholm P, *et al.* Effects of IKK inhibitor PS1145 on NF-kappaB function, proliferation, apoptosis and invasion activity in prostate carcinoma cells. Oncogene 2006; 25(3): 387-98.
[PMID: 16170348]

[133] Shivapurkar N, Reddy J, Chaudhary PM, Gazdar AF. Apoptosis and lung cancer: a review. J Cell Biochem 2003; 88(5): 885-98.
[http://dx.doi.org/10.1002/jcb.10440] [PMID: 12616528]

[134] Gupta P, Bansal MP, Koul A. Lycopene modulates initiation of *N*-nitrosodiethylamine induced hepatocarcinogenesis: studies on chromosomal abnormalities, membrane fluidity and antioxidant defense system. Chem Biol Interact 2013; 206(2): 364-74.

[http://dx.doi.org/10.1016/j.cbi.2013.10.010] [PMID: 24144777]

[135]  Bugianesi E. Non-alcoholic steatohepatitis and cancer. Clin Liver Dis 2007; 11(1): 191-207, x-xi.
[http://dx.doi.org/10.1016/j.cld.2007.02.006] [PMID: 17544979]

[136]  Mori S, Yamasaki T, Sakaida I, *et al.* Hepatocellular carcinoma with nonalcoholic steatohepatitis. J Gastroenterol 2004; 39(4): 391-6.
[http://dx.doi.org/10.1007/s00535-003-1308-3] [PMID: 15168253]

[137]  Wang Y, Ausman LM, Greenberg AS, Russell RM, Wang X-D. Dietary lycopene and tomato extract supplementations inhibit nonalcoholic steatohepatitis-promoted hepatocarcinogenesis in rats. Int J Cancer 2010; 126(8): 1788-96.
[PMID: 19551842]

[138]  Nwokocha CR, Nwokocha MI, Aneto I, *et al.* Comparative analysis on the effect of *Lycopersicon esculentum* (tomato) in reducing cadmium, mercury and lead accumulation in liver. Food Chem Toxicol 2012; 50(6): 2070-3.
[http://dx.doi.org/10.1016/j.fct.2012.03.079] [PMID: 22507840]

[139]  Tariq SA. Role of ascorbic acid in scavenging free radicals and lead toxicity from biosystems. Mol Biotechnol 2007; 37(1): 62-5.
[http://dx.doi.org/10.1007/s12033-007-0045-x] [PMID: 17914166]

[140]  Donpunha W, Kukongviriyapan U, Sompamit K, Pakdeechote P, Kukongviriyapan V, Pannangpetch P. Protective effect of ascorbic acid on cadmium-induced hypertension and vascular dysfunction in mice. Biometals 2011; 24(1): 105-15.
[http://dx.doi.org/10.1007/s10534-010-9379-0] [PMID: 20872046]

[141]  Pereira C, Barros L, Carvalho AM, Ferreira IC. Use of UFLC-PDA for the analysis of organic acids in thirty-five species of food and medicinal plants. Food Anal Methods 2013; 6: 1337-44.
[http://dx.doi.org/10.1007/s12161-012-9548-6]

[142]  Simion V, Câmpeanu GH, Vasile G, Artimon M, Catană L, Negoiţă M. Nitrate and nitrite accumulation in tomatoes and derived products. Rom Biotechnol Lett 2008; 13(4): 3785-90.

[143]  Kalogeropoulos N, Chiou A, Pyriochou V, Peristeraki A, Karathanos VT. Bioactive phytochemicals in industrial tomatoes and their processing byproducts. LWT - Food Sci Tech (Paris) 2012; 49(2): 213-6.

# CHAPTER 4

# Bioactive Compounds from *Capsicum annuum* as Health Promoters

**Adriana M. S. Sousa[1], Lorena Carro[2], Encarna Velàzquez[2], Carlos Albuquerque[3], Luís R. Silva[1,4,5,*]**

[1] *CICS UBI Health Sciences Research Centre, University of Beira Interior, 6201-506 Covilhã, Portugal*

[2] *Departamento de Microbiología y Genética, Facultad de Farmacia, Universidad de Salamanca, Salamanca, Spain*

[3] *IPV - ESSV Polytechnic Institute of Viseu, Higher Health School of Viseu, 3500-843, Viseu, Portugal*

[4] *IPCB ESALD Polytechnic Institute of Castelo Branco, Higher Health School Dr. Lopes Dias, 6000-767, Castelo Branco, Portugal*

[5] *LEPABE Department of Chemical Engineering, Faculty of Engineering, University of Porto, 4200-465 Porto, Portugal*

[6] *Mountain Research Centre (CIMO), ESA, Polytechnic Institute of Bragança, Campus de Santa Apolónia, Ap. 1172, 5301-855 Bragança, Portugal*

**Abstract:** *Capsicum annuum* (Pepper) is an agricultural crop of the Solanaceae family which is important as a vegetable food (bell pepper), as a spice (chili pepper), and as a colorant (paprika). It is native to Mexico and Central America but it is consumed worldwide and used in a great variety of dishes depending on its texture, flavor and color. *C. annuum* is a good source of bioactive compounds, such as polyphenols (flavonoids and phenolic acids), capsaicinoids, capsinoids, carotenoids and vitamins, with well-known anti-oxidant and anti-inflammatory effects. Since the human body cannot produce many of these compounds, this crop assumes an important role in health protection, and might present new pharmacological solutions to several conditions. In this chapter, we aim to summarize the phytochemical composition of

---

[*] **Corresponding author Luís R. Silva:** CICS-UBI - Health Sciences Research Center, University of Beira Interior, Covilhã, Portugal; Tel: +351 275 329 077; Fax: +351 275 329 099; Email: luisfarmacognosia@gmail.com.

**Luís Rodrigues da Silva and Branca Maria Silva (Eds.)**

*C. annuum* and emphasize the major and most recent findings concerning health promoting benefits of its bioactive compounds (polyphenols, capsaicinoids, capsinoids, carotenoids, vitamins, and phytosterols), particularly those related with anti-oxidant, anti-nociceptive, anti-neoplastic, and anti-obesity properties.

**Keywords:** Antioxidant, Bioactive compounds, *Capsicum annuum*, Health.

## INTRODUCTION

*Capsicum annuum* (Pepper) is an agricultural crop of the Solanaceae family and it is important as a vegetable food (bell pepper), as a spice (chili pepper) and colorant (paprika) [1].

**Fig. (1).** *Capsicum annuum* plant.

The Sonanaceae family includes 90 genera and 2000 species. The five most cultivated are *Capsicumannuum* (the dominant species) (Fig. **1**), *C. baccatum*, *C. chinense*, *C. frutescens*, and *C. pubescens* [2]. *C. annuum* is native to Mexico and Central America, but it is consumed worldwide and used in a great variety of dishes depending on its texture, flavor and color [1].

Pepper fruits are collected in distinct stages of maturation (visually distinguished by their color), from the most immature (green peppers – (Fig. **2**)) to the most mature (red peppers) [1]. This variation on color is related to the pepper's capacity to synthesize carotenoids and retain chlorophyll [3], and therefore different maturity stages present different composition. Additionally, during ripening, the fruits not only undergo a transformation in color but also in aroma and texture [3].

**Fig. (2).** *Capsicum annuum* fruit.

Apart from its culinary uses, *C. annuum* is also a good source of bioactive compounds, such as polyphenols (flavonoids and phenolic acids), capsaicinoids, capsinoids, carotenoids and vitamins, with well-known anti-oxidant and anti-inflammatory effects [3]. Since the human body cannot produce many of these compounds, this crop assumes an important role in health protection, and might present new pharmacological solutions to several conditions. Moreover, some of these bioactive compounds have antimicrobial activity against bacteria and yeasts [4 - 8] and antitumoral activity [8]. On the other hand, inoculation with some bacteria can modify the bioactive compounds content of *C. annuum* [9, 10].

In this chapter, we aim to summarize the phytochemical composition of *C. annuum* and emphasize the major and most recent findings concerning health promoting benefits of its bioactive compounds (polyphenols, capsaicinoids, capsinoids, carotenoids, vitamins, and phytosterols), especially those related to its anti-oxidant, anti-nociceptive, anti-neoplasic and anti-obesity properties.

## POLYPHENOLS

Polyphenols are an important group of secondary metabolites, synthesized by plants as a result of adaptation to stress conditions, such as infection, water/cold stress and wounding, among others [11]. These are some of the most abundant bioactive compounds present in peppers, and contribute to their color and flavor [2].

Polyphenols are divided into two main groups: phenolic acids and flavonoids (the most studied group). Flavonoid levels vary depending on several conditions such as genetic, developmental, and environmental [2, 11].

The main polyphenols in *C. annuum* are represented in Table **1**.

Table 1. Main polyphenols present in *C. annuum* [8].

| Compounds | Concentration (mg/g) |
|---|---|
| Quercetin rhamnoside | 82.6 |
| Luteolin glucoside | 35.46 |
| Quercetin glucoside | 19.86 |
| Kaempferol di-glucoside | 17.17 |
| Daphnetine | 16.29 |
| Quercetin | 10.81 |
| *p*-Coumaric acid | 6.97 |
| *p*-Coumaryl tyrosine | 6.81 |
| *p*-Coumaroyl glycolic | 6.47 |
| Luteolin di-glucoside | 5.66 |
| Vanilic acid glucoside | 4.02 |
| Hydroxybenzoylhexose | 3.29 |
| Hydrocaffeic acid | 3.03 |
| Caffeoyl glucoside | 2.59 |
| Hydroxycoumarin | 2.42 |
| Luteolin | 0.88 |

Phenolic compounds are mainly known and studied because of their anti-oxidant properties. During the normal metabolism of aerobic cells, free radicals are naturally produced as by-products. Polyphenols, as other antioxidants, possess free radical scavenging properties, thus protecting the human body against oxidative damage [11]. Several epidemiological studies have demonstrated that polyphenols might be involved in the prevention of a multitude of chronic diseases, such as cancer, cardiovascular diseases, and neurodegenerative diseases [11 - 13].

This antioxidant effect might be due to the presence of these reducing polyphenols in plasma after polyphenol rich-food, their effect upon concentration of other reducing agents, or their effect on the absorption of pro-oxidative compounds (*e.g.* Fe(II)) [13, 14]. In terms of cardiovascular diseases, polyphenols exert their protective role mainly because of their capacity to inhibit LDL (low-density lipoprotein) oxidation, which is the key mechanism to the formation of atherogenic plaques [13].

In terms of anti-cancer effects, several action mechanisms have been suggested such as anti-inflamatory activity, prevention of oxidation, induction of detoxification enzymes, antiproliferation, induction of apoptosis, regulation of the immune systems, among others [13, 15].

Additionally, these compounds inhibit pancreatic lipoprotein lipase (which transforms triglycerides into intestine-absorbable glycerol and fatty acids), participating in the prevention of obesity [12] and are also reported to have antimicrobial activity [7, 8].

## CAPSAICINOIDS

Capsaicinoids are a group of alkaloid compounds unique to chili peppers that are responsible for their pungent sensation [16]. These are synthesized by the condensation of fatty acids and vanillylamine and are mainly located in the tissue of the placenta [17]. Their concentration depends on the fruit genotype, maturity and conditions of cultivation [17].

The two major capsaicinoids are capsaicin (the main active compound of chili peppers) and dihydrocapsaicin, responsible for up to 90% of the pungency, followed by nordihydrocapsaicin, homodihydrocapsaicin, and homocapsaicin

[2, 17].

The chemical structure of capsaicin is represented in Fig. (**3**).

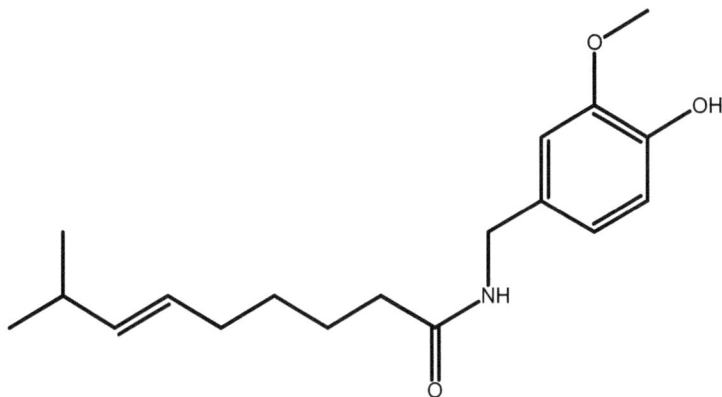

**Fig. (3).** Capsacin chemical structure.

Capsaicin is a selective agonist for the transient receptor potential vanilloid 1 (TRPV1) cation channel, present in sensory neurons as well as non-neural tissues [18].

The present clinical application of capsaicinoids is mainly pain management [19]. However, several studies have focused on other potential health benefits of the consumption of this compound, namely anti-obesity, anti-cancer, among others.

**Pain Relief**

One of the main health benefits of capsaicinoids is their anti-nociceptive effect, reported since 1850 [20], where the most studied capsaicinoid is capsaicin [17].

Capsaicin has an interesting mechanism of action. As mentioned before, it is a TRPV1 agonist; this non-selective channel is present in the neuronal membrane of primary sensory (afferent) neurons and in their C-fibers that is activated by diverse stimuli (*e.g.* chemical substances), and release several neuropeptides such as P substance [17, 21]. Capsaicin intensively stimulates this receptor, triggering the influx of calcium in to the neuron, with release of inflammatory mediators leading to pain, burning sensation, dermal irritation, among others [17, 21].

However, after this initial excitation of the nociceptive neurons, a long lasting refractory state occurs and these previously excited neurons become resistant various stimuli including pain [17, 19].

Pain treatments based on capsaicin take two forms: a low (0,025-0,075%) and a high dose (8%), both commercially available [21]. As a low dose form, it is added to several creams, with a poor to moderate efficacy for chronic pain [17]. As a high dose form, a 8% dermal patch has been approved by the U.S. Food and Drug Administration (FDA) for postherpetic neuralgia. It is intended to be applied for one hour without need to reapply for three months. Although it causes significant painful sensations during treatment and therefore a local anesthetic has to be given prior to the treatment, this high dose patch requires a single application for a large period of time. Recent studies have also shown promising results of this treatment on other pain syndromes such as post-surgical neuralgia, post-traumati--neuropathy, and mixed pain syndrome, as well as HIV polyneuropathy [22, 23]. Therefore, there is a need for studies focusing on this pain relieving potential on other chronic pain conditions.

## Anti-Obesity

Another important area where capsaicinoids have been widely studied is obesity, a growing concern in the last decade that poses a threat to public health [16].

The basis for weight management is based on two principles: the increase of energy expenditure and the reduction of intake [16]. Several animal studies [24 - 26] and clinical trials [26, 27] have found that prior consumption of capsaicinoids (*e.g.* in capsaicinoid rich red pepper, or capsaicin supplements) leads to appetite alterations, increase in energy expenditure, lipid oxidation and restriction in lipidogenesis.

The mechanism of the alterations in terms of appetite (such as reduction in hunger and desire to eat, as well as increase in society) is not yet understood [16]. Nonetheless, the increase in energy expenditure is caused by the increase of metabolic rate, body temperature and oxygen consumption. This is explained by the activation of TRPV1 channel, which leads to the release of catecholamines, stimulating the sympathetic nervous system [16]. The increase of plasmatic catecholamines also leads to lipid oxidation either by increasing free fatty acid levels or muscle tryglicerides lipolysis [27]. A proteomic approach was used in order to understand the molecular mechanisms of these events, and results from this study suggested that thermogenesis and lipid metabolism are markedly altered upon capsaicin treatment, supporting the important role of this compound in regulating energy metabolism [28].

Additionally, there are also studies focusing on obesity related diseases such as insulin resistance and type II diabetes. Studies performed with obese mice have shown that supplements of (or even topical) low doses of capsaicin lead to a reduction in fasting glucose and plasma triglyceride levels, and an increase in insulin sensitivity [29 - 31]. However, the chronic consumption of capsaicinoids for anti-obesity effects is limited by their strong pungency. Thus, capsinoids, the non-pungent analogs of capsaicinoids, present an interesting alternate solution [17].

## Anti-Cancer

Capsaicinoids also present anti-inflammatory, antioxidant, antiproliferative and anti-cancer potential [32, 33]. Studies have already demonstrated its capability to mediate cell cycle arrest and induce apoptosis *in vitro* in several cancer cell lines, such as colon carcinoma [34], prostate cancer [35], gastric cancer [36], nasopharyngeal carcinoma [37], hepatocarcinoma [38], breast cancer [39], glioma [40], multiple myeloma [41], tongue cancer [42], esophageal carcinoma [43], pancreatic cancer [44], and non-small cell lung cancer [45], as well as human leukemic cells [46].

Nonetheless, conflicting data has also shown that capsaicin in certain circumstances might also have the potential for carcinogenesis [47]. Therefore, more research is needed in order to understand the role of capsaicinoids in terms of cancer genesis/prevention.

## Other Effects

In addition to the previously mentioned effects, studies have also shown the potential of capsaicinoids against cardiovascular [19] and gastrointestinal diseases

[19] as well as antimicrobial [6] and antiviral agents [48].

## CAPSINOIDS

Capsinoids are the non-pungent analogs of capsacinoids and are found on (CH-19) sweet peppers [19]. These consist mainly of capsiate, dihydrocapsiate and norhydrocapsiate, which are also TRPV1 agonists, and activators of the sympathetic nervous system [19]. However, these compounds are unable to get to the TRPV1 located in the mouth, thus not inducing the pungent sensation [17]. It is also reported that these compounds might present anti-obesity and anti-cancer properties

The chemical structure of capsiate is represented in Fig. (**4**).

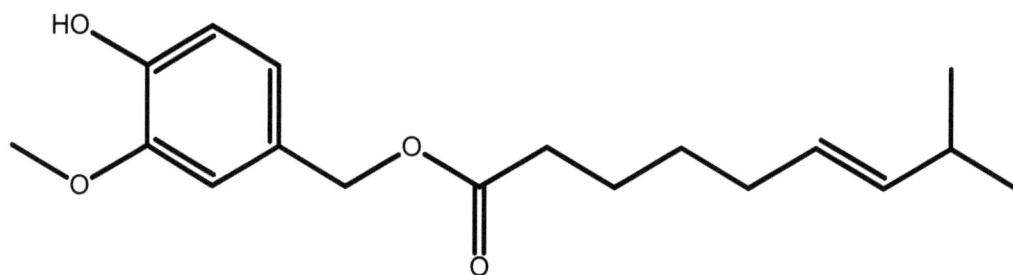

**Fig. (4).** Capsiate chemical structure.

## Anti-Obesity

As TRPV1 agonists and sympathetic nervous system activators, capsinoids have also been shown to increase body temperature and oxygen consumption, as well as increase energy expenditure and inhibit lipid accumulation [49], thus contributing to weight loss.

Because of their lack of pungency, these compounds might present a much more interesting option when compared to capsaicinoids [17].

## Anti-Cancer

Although it has been shown capsinoid capability of anti-cancer and chemopreventive activities [50], limited literature is available so far [17].

## CAROTENOIDS

*C. annuum* fruits are collected in distinct stages of maturation (distinguished by their color), from the most immature (green peppers) to the most mature (red peppers) [1]. This variation on color is related to the pepper's capacity to synthesize carotenoids and retain chlorophyll [3].

Chlorophyll and carotenoids typical of the chloroplasts contribute to the green colored pepper; $\alpha$- and $\beta$-carotene, zeaxanthin, lutein and $\beta$-cryptoxanthin to the yellow-orange color; and carotenoid pigments of capsanthin, capsorubin and capsanthin 5,6-epoxide to the red color [3]. The color is a crucial factor in terms of purchasing decision for the consumer, being green bell pepper the most consumed and produced [51].

This crop is a major source of carotenoids. Carotenoids are important as precursors of Vitamin A ($\beta$-carotene, $\alpha$-carotene, and $\beta$-cryptoxanthin), since the human body is not able to synthesize it *de novo* from endogenous precursors [52], but also as antioxidants, protecting cells against oxidative stress, and thus protectors against several diseases such as cancer, cardiovascular diseases, as well as age-related macular degeneration and cataracts [52, 53].

## VITAMINS

### Ascorbic Acid

One of the main nutritional qualities of *C. annuum* is its content in ascorbic acid (vitamin C), as it is one of the vegetables with a higher content [54]. This is of major importance since the human body is not able to synthesize it [2].

Levels of Vitamin C in this crop depend on factors such as cultivar, maturity at harvest (its levels increase during ripening), production practices and storage conditions [54]. Furthermore, cooking processes, such as boiling and steaming, especially when prolonged, severely reduce its levels [55].

Vitamin C content is a necessary nutrient to human body, and it's mainly associated with antioxidant activities thus involved in the prevention of degenerative diseases, certain types of cancer, as well as cardiovascular diseases, hypertension and hypercholesterolemia [53].

## PHYTOSTEROLS

Phytosterols (plant sterols) are cholesterol-like bioactive compounds present in the cell membranes of vegetable foods [56]. They may be present as free or esterified sterols, such as β-sitosterol, campesterol and stigmasterol [56]. The chemical structures of the aforementioned sterols are represented in Figs. (**5-7**).

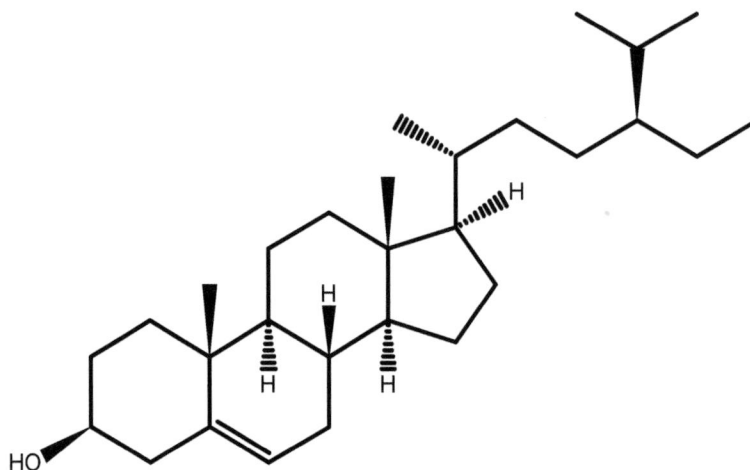

**Fig. (5).** *β*-Sitosterol chemical structure.

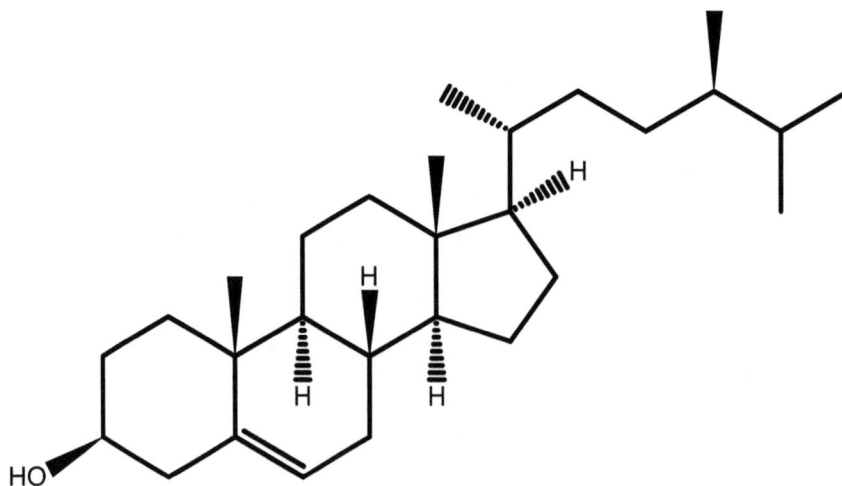

**Fig. (6).** Campesterol chemical structure.

*C. annuum* contains mostly sitosterol (52-74% of total phytosterol content), followed by campesterol (21-27%), and stigmasterol (minor sterol component) [57], and their content decreases during ripening [58].

These compounds are known for their hypocholesterolemic properties associated with intestinal cholesterol absortion reduction [56, 59], thus reducing coronary risk. Also, phytosterols are reported to be protective against colon cancer as well as benign prostatic hyperplasia, to retard the growth and metastization of breast cancer cells [60], and to modulate immune system activity [61].

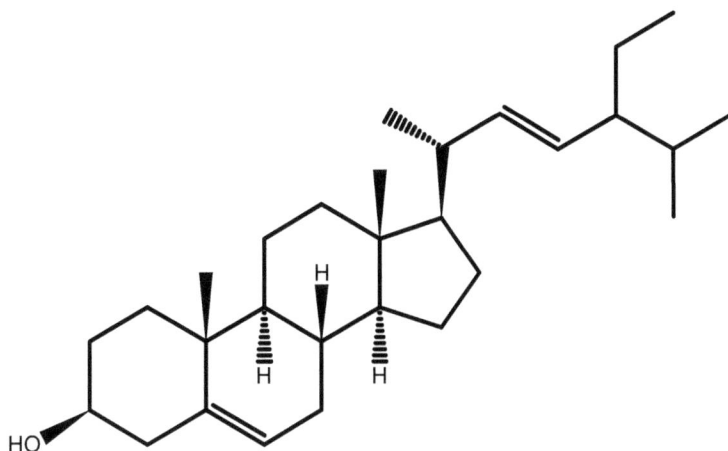

**Fig. (7).** Stigmasterol chemical structure.

## CONCLUDING REMARKS

*C. annuum* is a very rich vegetable in terms of bioactive compounds, such as polyphenols, capsaicinoids and capsinoids, as well as carotenoids, vitamins and phytosterols. Several of these compounds have been associated with great health benefits, namely antioxidant effects, with the potential for the prevention of several diseases such as cancer, cardiovascular and neurodegenerative diseases, as well as macular diseases. However, there is a need for extensive research on some of the bioactive compounds, especially those unique to peppers, such as capsaicinoids and capsinoids, the promising non-pungent analogs of capsainoids.

## CONFLICT OF INTEREST

The authors confirm that they have no conflict of interest to declare for this publication.

## ACKNOWLEDGEMENTS

This work was supported by the Portuguese "Fundação para a Ciência e a Tecnologia" - FCT: L.R. Silva (SFRH/BPD/105263/2014; CICS-UBI); UMIB (Pest-OE/SAU/UI0215/2014), co-funded by FEDER *via* Programa Operacional Fatores de Competitividade - COMPETE/QREN & FSE and POPH funds and Ministerio de Ciencia e Innovación (MINECO) and Junta de Castilla y León from Spain. Additionally, the authors would like to thank our numerous collaborators and students involved in this research over the years.

## REFERENCES

[1]   Silva LR, Azevedo J, Pereira MJ, Valentão P, Andrade PB. Chemical assessment and antioxidant capacity of pepper (Capsicum annuum L.) seeds. Food Chem Toxicol 2013; 53: 240-8.
      [http://dx.doi.org/10.1016/j.fct.2012.11.036] [PMID: 23238236]

[2]   Jayaprakasha GK, Haejin B, Kevin C, John LJ, Bhimanagouda SP. Bioactive compounds in peppers and their antioxidant potential. In: Tunick M, Mejía E, Eds. Hispanic foods: chemistry and bioactive compounds. USA: American Chemical Society 2012.
      [http://dx.doi.org/10.1021/bk-2012-1109.ch004]

[3]   Sun T, Xu Z, Wu CT, Janes M, Prinyawiwatkul W, No HK. Antioxidant activities of different colored sweet bell peppers (*Capsicum annuum* L.). J Food Sci 2007; 72(2): S98-S102.
      [http://dx.doi.org/10.1111/j.1750-3841.2006.00245.x] [PMID: 17995862]

[4]   De Marino S, Borbone N, Gala F, *et al.* New constituents of sweet *Capsicum annuum* L. fruits and evaluation of their biological activity. J Agric Food Chem 2006; 54(20): 7508-16.
      [http://dx.doi.org/10.1021/jf061404z] [PMID: 17002415]

[5]   Nazzaro F, Caliendo G, Arnesi G, Veronesi A, Sarzi P, Fratianni F. Comparative content of some bioactive compounds in two varieties of Capsicum annuum L. sweet pepper and evaluation of their antimicrobial and mutagenic activities. J Food Biochem 2009; 33: 852-68.
      [http://dx.doi.org/10.1111/j.1745-4514.2009.00259.x]

[6]   Santos MM, Vieira-da-Motta O, Vieira IJ, *et al.* Antibacterial activity of *Capsicum annuum* extract and synthetic capsaicinoid derivatives against *Streptococcus mutans*. J Nat Med 2012; 66(2): 354-6.
      [http://dx.doi.org/10.1007/s11418-011-0579-x] [PMID: 21858615]

[7]   El Ksibi I, Ben Slama R, Faidi K, Ben Ticha M, M'henni MF. Mixture approach for optimizing the recovery of colored phenolics from red pepper (Capsicum annum L.) by-products as potential source of natural dye and assessment of its antimicrobial activity. Ind Crops Prod 2015; 70: 34-40.

[http://dx.doi.org/10.1016/j.indcrop.2015.03.017]

[8]     Mokhtar M, Soukup J, Donato P, *et al.* Determination of the polyphenolic content of a Capsicum annuum L. extract by liquid chromatography coupled to photodiode array and mass spectrometry detection and evaluation of its biological activity. J Sep Sci 2015; 38(2): 171-8.
[http://dx.doi.org/10.1002/jssc.201400993] [PMID: 25378270]

[9]     Silva LR, Azevedo J, Pereira MJ, *et al.* Inoculation of the nonlegume *Capsicum annuum* (L.) with *Rhizobium* strains. 1. Effect on bioactive compounds, antioxidant activity, and fruit ripeness. J Agric Food Chem 2014; 62(3): 557-64.
[http://dx.doi.org/10.1021/jf4046649] [PMID: 24404842]

[10]    Silva LR, Azevedo J, Pereira MJ, *et al.* Inoculation of the nonlegume *Capsicum annuum* L. with *Rhizobium* strains. 2. Changes in sterols, triterpenes, fatty acids, and volatile compounds. J Agric Food Chem 2014; 62(3): 565-73.
[http://dx.doi.org/10.1021/jf4046655] [PMID: 24405510]

[11]    Materska M, Perucka I. Antioxidant activity of the main phenolic compounds isolated from hot pepper fruit (*Capsicum annuum* L). J Agric Food Chem 2005; 53(5): 1750-6.
[http://dx.doi.org/10.1021/jf035331k] [PMID: 15740069]

[12]    Jeong WY, Jin JS, Cho YA, *et al.* Determination of polyphenols in three *Capsicum annuum* L. (bell pepper) varieties using high-performance liquid chromatography-tandem mass spectrometry: their contribution to overall antioxidant and anticancer activity. J Sep Sci 2011; 34(21): 2967-74.
[http://dx.doi.org/10.1002/jssc.201100524] [PMID: 21898818]

[13]    Pandey KB, Rizvi SI. Plant polyphenols as dietary antioxidants in human health and disease. Oxid Med Cell Longev 2009; 2(5): 270-8.
[http://dx.doi.org/10.4161/oxim.2.5.9498] [PMID: 20716914]

[14]    Oboh G, Rocha JB. Polyphenols in red pepper [Capsicum annuum var. aviculare (Tepin)] and their protective effect on some pro-oxidants induced lipid peroxidation in brain and liver. Eur Food Res Technol 2007; 225: 239-47.
[http://dx.doi.org/10.1007/s00217-006-0410-1]

[15]    Scalbert A, Manach C, Morand C, Rémésy C, Jiménez L. Dietary polyphenols and the prevention of diseases. Crit Rev Food Sci Nutr 2005; 45(4): 287-306.
[http://dx.doi.org/10.1080/1040869059096] [PMID: 16047496]

[16]    Whiting S, Derbyshire E, Tiwari BK. Capsaicinoids and capsinoids. A potential role for weight management? A systematic review of the evidence. Appetite 2012; 59(2): 341-8.
[http://dx.doi.org/10.1016/j.appet.2012.05.015] [PMID: 22634197]

[17]    Luo XJ, Peng J, Li YJ. Recent advances in the study on capsaicinoids and capsinoids. Eur J Pharmacol 2011; 650(1): 1-7.
[http://dx.doi.org/10.1016/j.ejphar.2010.09.074] [PMID: 20946891]

[18]    Yang D, Luo Z, Ma S, *et al.* Activation of TRPV1 by dietary capsaicin improves endothelium-dependent vasorelaxation and prevents hypertension. Cell Metab 2010; 12(2): 130-41.
[http://dx.doi.org/10.1016/j.cmet.2010.05.015] [PMID: 20674858]

[19]    Sharma SK, Vij AS, Sharma M. Mechanisms and clinical uses of capsaicin. Eur J Pharmacol 2013;

720(1-3): 55-62.
[http://dx.doi.org/10.1016/j.ejphar.2013.10.053] [PMID: 24211679]

[20]    Anand P, Bley K. Topical capsaicin for pain management: therapeutic potential and mechanisms of action of the new high-concentration capsaicin 8% patch. Br J Anaesth 2011; 107(4): 490-502.
[http://dx.doi.org/10.1093/bja/aer260] [PMID: 21852280]

[21]    Smith H, Brooks JR. Capsaicin-based therapies for pain control. AbdelSalam O, Ed capsaicin as a therapeutic molecule. Basel: Springer 2014.
[http://dx.doi.org/10.1007/978-3-0348-0828-6_5]

[22]    Maihofner C, Heskamp M-L. Prospective, non-interventional study on the tolerability and analgesic effectiveness over 12 weeks after a single application of capsaicin 8% cutaneous patch in 1044 patients with peripheral neuropathic pain: first results of the QUEPP study. Curr Med Res Opin 2013; 29(6): 673-83.
[http://dx.doi.org/10.1185/03007995.2013.792246] [PMID: 23551064]

[23]    Simpson DM, Brown S, Tobias JK, Vanhove GF. NGX-4010, a capsaicin 8% dermal patch, for the treatment of painful HIV-associated distal sensory polyneuropathy: results of a 52-week open-label study. Clin J Pain 2014; 30(2): 134-42.
[PMID: 23446088]

[24]    Hsu CL, Yen GC. Effects of capsaicin on induction of apoptosis and inhibition of adipogenesis in 3T3-L1 cells. J Agric Food Chem 2007; 55(5): 1730-6.
[http://dx.doi.org/10.1021/jf062912b] [PMID: 17295509]

[25]    Kawada T, Hagihara K, Iwai K. Effects of capsaicin on lipid metabolism in rats fed a high fat diet. J Nutr 1986; 116(7): 1272-8.
[PMID: 2875141]

[26]    Reinbach HC, Smeets A, Martinussen T, Møller P, Westerterp-Plantenga MS. Effects of capsaicin, green tea and CH-19 sweet pepper on appetite and energy intake in humans in negative and positive energy balance. Clin Nutr 2009; 28(3): 260-5.
[http://dx.doi.org/10.1016/j.clnu.2009.01.010] [PMID: 19345452]

[27]    Shin KO, Moritani T. Alterations of autonomic nervous activity and energy metabolism by capsaicin ingestion during aerobic exercise in healthy men. J Nutr Sci Vitaminol (Tokyo) 2007; 53(2): 124-32.
[http://dx.doi.org/10.3177/jnsv.53.124] [PMID: 17615999]

[28]    Joo JI, Kim DH, Choi JW, Yun JW. Proteomic analysis for antiobesity potential of capsaicin on white adipose tissue in rats fed with a high fat diet. J Proteome Res 2010; 9(6): 2977-87.
[http://dx.doi.org/10.1021/pr901175w] [PMID: 20359164]

[29]    Kang JH, Tsuyoshi G, Le Ngoc H, *et al.* Dietary capsaicin attenuates metabolic dysregulation in genetically obese diabetic mice. J Med Food 2011; 14(3): 310-5.
[http://dx.doi.org/10.1089/jmf.2010.1367] [PMID: 21332406]

[30]    Lee GR, Shin MK, Yoon DJ, *et al.* Topical application of capsaicin reduces visceral adipose fat by affecting adipokine levels in high-fat diet-induced obese mice. Obesity (Silver Spring) 2013; 21(1): 115-22.
[http://dx.doi.org/10.1002/oby.20246] [PMID: 23505175]

[31]   Okumura T, Tsukui T, Hosokawa M, Miyashita K. Effect of caffeine and capsaicin on the blood glucose levels of obese/diabetic KK-A(y) mice. J Oleo Sci 2012; 61(9): 515-23.
[http://dx.doi.org/10.5650/jos.61.515] [PMID: 22975786]

[32]   Lin CH, Lu WC, Wang CW, Chan YC, Chen MK. Capsaicin induces cell cycle arrest and apoptosis in human KB cancer cells. BMC Complement Altern Med 2013; 13: 46.
[http://dx.doi.org/10.1186/1472-6882-13-46] [PMID: 23433093]

[33]   Oyagbemi AA, Saba AB, Azeez OI. Capsaicin: a novel chemopreventive molecule and its underlying molecular mechanisms of action. Indian J Cancer 2010; 47(1): 53-8.
[http://dx.doi.org/10.4103/0019-509X.58860] [PMID: 20071791]

[34]   Jin J, Lin G, Huang H, *et al.* Capsaicin mediates cell cycle arrest and apoptosis in human colon cancer cells *via* stabilizing and activating p53. Int J Biol Sci 2014; 10(3): 285-95.
[http://dx.doi.org/10.7150/ijbs.7730] [PMID: 24643130]

[35]   Mori A, Lehmann S, O'Kelly J, *et al.* Capsaicin, a component of red peppers, inhibits the growth of androgen-independent, p53 mutant prostate cancer cells. Cancer Res 2006; 66(6): 3222-9.
[http://dx.doi.org/10.1158/0008-5472.CAN-05-0087] [PMID: 16540674]

[36]   Huh HC, Lee SY, Lee SK, Park NH, Han IS. Capsaicin induces apoptosis of cisplatin-resistant stomach cancer cells by causing degradation of cisplatin-inducible Aurora-A protein. Nutr Cancer 2011; 63(7): 1095-103.
[http://dx.doi.org/10.1080/01635581.2011.607548] [PMID: 21932983]

[37]   Ip SW, Lan SH, Lu HF, *et al.* Capsaicin mediates apoptosis in human nasopharyngeal carcinoma NPC-TW 039 cells through mitochondrial depolarization and endoplasmic reticulum stress. Hum Exp Toxicol 2012; 31(6): 539-49.
[http://dx.doi.org/10.1177/0960327111417269] [PMID: 21859781]

[38]   Huang SP, Chen JC, Wu CC, *et al.* Capsaicin-induced apoptosis in human hepatoma HepG2 cells. Anticancer Res 2009; 29(1): 165-74.
[PMID: 19331147]

[39]   Chou CC, Wu YC, Wang YF, Chou MJ, Kuo SJ, Chen DR. Capsaicin-induced apoptosis in human breast cancer MCF-7 cells through caspase-independent pathway. Oncol Rep 2009; 21(3): 665-71.
[PMID: 19212624]

[40]   Amantini C, Mosca M, Nabissi M, *et al.* Capsaicin-induced apoptosis of glioma cells is mediated by TRPV1 vanilloid receptor and requires p38 MAPK activation. J Neurochem 2007; 102(3): 977-90.
[http://dx.doi.org/10.1111/j.1471-4159.2007.04582.x] [PMID: 17442041]

[41]   Bhutani M, Pathak AK, Nair AS, *et al.* Capsaicin is a novel blocker of constitutive and interleukin--inducible STAT3 activation. Clin Cancer Res 2007; 13(10): 3024-32.
[http://dx.doi.org/10.1158/1078-0432.CCR-06-2575] [PMID: 17505005]

[42]   Ip SW, Lan SH, Huang AC, *et al.* Capsaicin induces apoptosis in SCC-4 human tongue cancer cells through mitochondria-dependent and -independent pathways. Environ Toxicol 2012; 27(6): 332-41.
[http://dx.doi.org/10.1002/tox.20646] [PMID: 20925121]

[43]   Wu CC, Lin JP, Yang JS, *et al.* Capsaicin induced cell cycle arrest and apoptosis in human esophagus epidermoid carcinoma CE 81T/VGH cells through the elevation of intracellular reactive oxygen

species and Ca2+ productions and caspase-3 activation. Mutat Res 2006; 601(1-2): 71-82.
[http://dx.doi.org/10.1016/j.mrfmmm.2006.06.015] [PMID: 16942782]

[44]    Pramanik KC, Boreddy SR, Srivastava SK. Role of mitochondrial electron transport chain complexes in capsaicin mediated oxidative stress leading to apoptosis in pancreatic cancer cells. PLoS One 2011; 6(5): e20151.
[http://dx.doi.org/10.1371/journal.pone.0020151] [PMID: 21647434]

[45]    Brown KC, Witte TR, Hardman WE, *et al.* Capsaicin displays anti-proliferative activity against human small cell lung cancer in cell culture and nude mice models *via* the E2F pathway. PLoS One 2010; 5(4): e10243.
[http://dx.doi.org/10.1371/journal.pone.0010243] [PMID: 20421925]

[46]    Ito K, Nakazato T, Yamato K, *et al.* Induction of apoptosis in leukemic cells by homovanillic acid derivative, capsaicin, through oxidative stress: implication of phosphorylation of p53 at Ser-15 residue by reactive oxygen species. Cancer Res 2004; 64(3): 1071-8.
[http://dx.doi.org/10.1158/0008-5472.CAN-03-1670] [PMID: 14871840]

[47]    Surh YJ, Lee SS. Capsaicin in hot chili pepper: carcinogen, co-carcinogen or anticarcinogen? Food Chem Toxicol 1996; 34(3): 313-6.
[http://dx.doi.org/10.1016/0278-6915(95)00108-5] [PMID: 8621114]

[48]    Khan FA, Mahmood T, Ali M, Saeed A, Maalik A. Pharmacological importance of an ethnobotanical plant: Capsicum *annuum* L. Nat Prod Res 2014; 28(16): 1267-74.
[http://dx.doi.org/10.1080/14786419.2014.895723] [PMID: 24650229]

[49]    Faraut B, Giannesini B, Matarazzo V, *et al.* Capsiate administration results in an uncoupling protein-3 downregulation, an enhanced muscle oxidative capacity and a decreased abdominal fat content *in vivo.* Int J Obes 2009; 33(12): 1348-55.
[http://dx.doi.org/10.1038/ijo.2009.182] [PMID: 19773740]

[50]    Pyun BJ, Choi S, Lee Y, *et al.* Capsiate, a nonpungent capsaicin-like compound, inhibits angiogenesis and vascular permeability *via* a direct inhibition of Src kinase activity. Cancer Res 2008; 68(1): 227-35.
[http://dx.doi.org/10.1158/0008-5472.CAN-07-2799] [PMID: 18172315]

[51]    Frank CA, Nelson RG, Simonne EH, Behe BK, Simonne AE. Consumer preferences for color, price, and vitamin c content of bell peppers. HortScience 2001; 36: 795-800.

[52]    Gómez-García MdelR, Ochoa-Alejo N. Biochemistry and molecular biology of carotenoid biosynthesis in chili peppers (Capsicum spp.). Int J Mol Sci 2013; 14(9): 19025-53.
[http://dx.doi.org/10.3390/ijms140919025] [PMID: 24065101]

[53]    Marín A, Ferreres F, Tomás-Barberán FA, Gil MI. Characterization and quantitation of antioxidant constituents of sweet pepper (*Capsicum annuum* L.). J Agric Food Chem 2004; 52(12): 3861-9.
[http://dx.doi.org/10.1021/jf0497915] [PMID: 15186108]

[54]    Vanderslice JT, Higgs DJ, Hayes JM, Block G. Ascorbic acid and dehydroascorbic acid content of foods-as-eaten. J Food Compos Anal 1990; 3: 105-18.
[http://dx.doi.org/10.1016/0889-1575(90)90018-H]

[55]    Hwang IG, Shin YJ, Lee S, Lee J, Yoo SM. Effects of different cooking methods on the antioxidant

properties of red pepper (*capsicum annuum* L.). Prev Nutr Food Sci 2012; 17(4): 286-92.
[http://dx.doi.org/10.3746/pnf.2012.17.4.286] [PMID: 24471098]

[56]   Piironen V, Lindsay DG, Miettinen TA, Toivo J, Lampi A-M. Plant sterols: biosynthesis, biological function and their importance to human nutrition. J Sci Food Agric 2000; 80: 939-66.
[http://dx.doi.org/10.1002/(SICI)1097-0010(20000515)80:7<939::AID-JSFA644>3.0.CO;2-C]

[57]   Whitaker B. Steryl lipid content and composition in bell pepper fruit at three stages of ripening. J Am Soc Hortic Sci 1989; 114(4): 648-51.

[58]   Bhandari SR, Jung B-D, Baek H-Y, Lee Y-S. Ripening-dependent changes in phytonutrients and antioxidant activity of red pepper (*capsicum annuum* L.) fruits cultivated under open-field conditions. HortScience 2013; 48: 1275-82.

[59]   Han JH, Yang YX, Feng MY. Contents of phytosterols in vegetables and fruits commonly consumed in China. Biomed Environ Sci 2008; 21(6): 449-53.
[http://dx.doi.org/10.1016/S0895-3988(09)60001-5] [PMID: 19263798]

[60]   Awad AB, Downie A, Fink CS, Kim U. Dietary phytosterol inhibits the growth and metastasis of MDA-MB-231 human breast cancer cells grown in SCID mice. Anticancer Res 2000; 20(2A): 821-4.
[PMID: 10810360]

[61]   QuIlez J, GarcIa-Lorda P, Salas-Salvadó J. Potential uses and benefits of phytosterols in diet: present situation and future directions. Clin Nutr 2003; 22(4): 343-51.
[http://dx.doi.org/10.1016/S0261-5614(03)00060-8] [PMID: 12880600]

# CHAPTER 5

# Phytochemical, Nutritional, Antioxidant and Anticancer Properties of *Juglans regia* (L.)

**Branca M. Silva[1,\*], Ana R. Nunes[1], Elsa Ramalhosa[2], José A. Pereira[2], Marco G. Alves[1], Pedro F. Oliveira[1,3]**

[1] *CICS – UBI – Health Sciences Research Centre, University of Beira Interior, 6201-506 Covilhã, Portugal*

[2] *Mountain Research Centre (CIMO), School of Agriculture, Polytechnic Institute of Bragança, Campus de Sta Apolónia, Apartado 1172, 5301-855 Bragança, Portugal*

[3] *Department of Microscopy, Laboratory of Cell Biology, Institute of Biomedical Sciences Abel Salazar (ICBAS) and Unit for Multidisciplinary Research in Biomedicine (UMIB), University of Porto, 4050-313 Porto, Portugal*

**Abstract:** *Juglans regia* (L.) is a deciduous tree, spread throughout the world due to its nutritional and sensory value. Although walnuts, the *J. regia* seeds, have been part of the human diet for a long time, only in the recent years several studies have surfaced on this botanical species as a rich and inexpensive source of bioactive compounds. In fact, much research is available depicting the human health benefits of *J. regia* seeds, leaves and green husks for a wide variety of diseases mediated by oxidative stress, namely cancer, and neurodegenerative disorders. Among bioactive compounds, polyphenols play an important role. They are well known for their strong antioxidant, antihemolytic, antidiabetic, antimicrobial, antimutagenic and anticarcinogenic activities and are considered by many researchers responsible for the beneficial health effects of *J. regia* and its derivatives. In this chapter, we aim to emphasize the phytochemical composition, nutritional value and health promoting properties of *J. regia*, especially those related to its antioxidant and anticancer activities. Among several nut types, walnut stands up as a functional food, rich in antioxidants, especially in phenolic compounds and tocopherols. The potential for *J. regia* to be used as both preventive and therapeutic measure in several human diseases is high. For that reason, a lot of

---

\* **Corresponding author Branca M. Silva:** CICS – UBI – Health Sciences Research Centre, University of Beira Interior Av. Infante D. Henrique, 6201-506 Covilhã, Portugal; Tel: +351 275 329 077; Fax: +351 275 329 099; Email: bmcms@ubi.pt

attention has been given to *J. regia* in the last decade.

Our chapter, summarizes the main studies about *J. regia* and its beneficial effects on human health, focusing on its antioxidant and anticancer activities.

**Keywords:** Antioxidants, Cancer, *Juglans regia*, Juglone, Polyphenols, Walnut.

## INTRODUCTION

Walnut tree (*Juglans regia* L.) is the best known member of the *Juglans* genus, being regarded as a valuable crop all over the world [1]. This deciduous tree species, native to central Asia, the western Himalayan chain and Kyrgyzstan [2], is now cultivated around the World, namely in North and South America, North Africa, East Asia and Europe [3]. Part of the high worth attributed to *J. regia* is due to its nutritional and pharmaceutical properties, and commercial value of several of its components, particularly its leaves [1, 4], dry seeds [5], green husks [6] and alcoholic beverages produced from the green fruits [7]. The most widely used morphological part of *J. regia* is the seed, the walnut, and valued for its nutritional and sensorial attributes and, health benefits. Walnuts have been consumed since pre-agriculture times [8] mainly because they are a nutrient-dense food, rich in polyunsaturated fats and proteins [8, 9]. It can be consumed fresh or toasted [10]. The walnut also has a very interesting micronutrient profile, containing several vitamins and minerals [5]. If left in its shell, the walnut can be stored for a long time without losing its nutritional properties [9]. It has a slight astringent flavor, associated to the presence of phenolic compounds, which possess potent antioxidant properties [11, 12]. Moreover, walnuts present the highest antioxidant activity when compared with several other types of nuts [9], namely peanuts, pistachios, hazelnuts and almonds. Besides, walnut antiproliferative activity against human colon and kidney cancers has already been reported by Carvalho *et al.* [3].

*J. regia* leaves are not as popular as the seeds, however they are used in the preparation of infusions in several European countries, especially in rural areas [1]. Noteworthy, several beneficial effects have been attributed to these leaves such as antidiarrheic, antihelmintic, depurative, keratolytic, antifungal, hypoglycaemic, hypotensive, anti-scrofulous as well as sedative and astringent properties [13 - 17]. More recently, the antioxidant and antiproliferative effects of *J. regia* leaves have started to be studied with promising results [1, 3]. Walnut green husk (WGH), a waste from the walnut production, is produced in large quantity. Its use is scarce and the possible beneficial effects are not very explored. However, it has been recently reported that the methanolic extracts of WGH present high antioxidant and antiproliferative effects, comparable to the seeds and leaves [1, 3]. Thus, it is expected that these extracts may present great health benefits.

In this chapter we intend to discuss the recent findings concerning the phytochemical, nutritional, antioxidant and anticancer properties of *J. regia,* focusing in the seeds, leaves and green husks. Overall, we intend to present an up-to-date overview of the potential health beneficial effects of *J. regia.*

## CHEMICAL COMPOSITION AND NUTRITIONAL PROPERTIES

Walnut is one of the most popular nuts probably due to its attractive sensorial attributes and also to its health promoting properties. This is a nutrient dense food with high total caloric value and a significant content of important nutrients and bioactive phytocomponents with great relevance in human health. In fact, this nut possesses a complex composition containing hundreds of compounds, which belong to several different classes. Of note is that walnuts constitute a significant dietary source of polyunsaturated fats, proteins, fibers, melatonin, vitamin E, and polyphenols [1, 3, 9, 12, 18 - 20]. When compared with other types of nuts, walnuts are very rich in antioxidants, mainly phenolic compounds and tocopherols. A great part of polyphenols can be found in the pellicles [3, 9].

### Major Components

Due to walnut high lipid content and low moisture, it presents a high total caloric content of approximately 700 Kcal / 100 g of edible portion [12]. Indeed, its major food components are lipids, followed by proteins (Fig. **1**) [12]. Other major components include fibers, water and carbohydrates, although present in much lower amounts (Fig. **1**) [12].

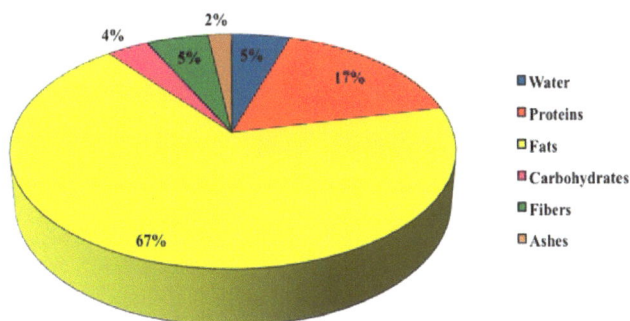

**Fig. (1).** Walnut chemical composition.

## *Macronutrients*

Walnuts are nutrient dense food products. Its lipid content is very high, being one of the plant foods richest in fat [21]. Pereira *et al.* [12] have determined the chemical profile of several cultivars (Franquette, Lara, Marbot, Mayette, Mellanaise and Parisiene cvs.), and found similar results concerning lipid contents. In fact, fat was always the major food component in all cultivars, with contents between 69 and 72% of edible portion (in cvs. Marbot and Franquette, respectively). The fatty acids profile was very similar in the six walnut cultivars [12] and was dominated by polyunsaturated fatty acids (PUFA) [12]. Linoleic acid, an essential PUFA, was the major walnut fatty acid [12] and was followed by oleic and linolenic acids [12]. Monounsaturated fatty acids (MUFA) were also present in considerable amounts and, saturated fatty acids (SFA) were the less abundant class in walnuts [12].

As far as we know, walnut proteins remain largely unstudied but according to Pereira *et al.* [12] the protein contents are between 14 and 18% in cvs. Lara and Marbot, respectively. Walnut total carbohydrates content is low. Accordingly, Pereira *et al.* [12] have found carbohydrates contents between 4 and 7% of edible portion (in cvs. Mayette and Lara, respectively). Monosaccharides and disaccharides classes are the most representative ones, but starch is also present in this nut [21]. Walnuts constitute a very significant source of fibers [21].

## Minor Components

### *Micronutrients*

Walnut contains several water-soluble and lipid-soluble vitamins, namely vitamins C, $B_1$, $B_2$, $B_3$, $B_6$, and E, and is considered a good source of this last vitamin [19, 20]. Vitamin E is composed by a group of minor fat-soluble compounds, including tocopherols ($\alpha$-, $\beta$-, $\gamma$-, and $\delta$-) and tocotrienols ($\alpha$-, $\beta$-, $\gamma$-, and $\delta$-) (Fig. **2**).

Vitamin E has strong antioxidant potential and play important biochemical and physiological roles. Amaral *et al.* [19] have determined the vitamin E profile of several walnut cultivars and found that the qualitative profile is very similar in all those varieties and composed by the four tocopherols and $\gamma$-tocotrienol, being $\gamma$-tocopherol the major one followed by $\alpha$- and $\delta$-tocopherols.

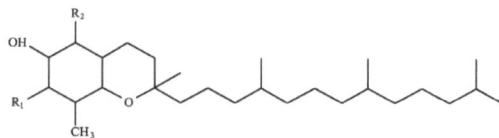

$\alpha$ - tocopherol: $R_1 = R_2 = CH_3$

$\beta$ - tocopherol: $R_1 = H$; $R_2 = CH_3$

$\delta$ - tocopherol: $R_1 = CH_3$ ; $R_2 = H$

$\gamma$ - tocopherol: $R_1 = R_2 = H$

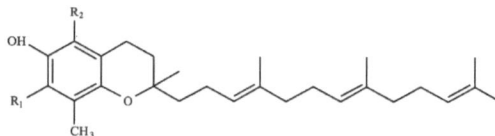

$\alpha$ - tocotrienol: $R_1 = R_2 = CH_3$

$\beta$ - tocotrienol: $R_1 = H$; $R_2 = CH_3$

$\delta$ - tocotrienol: $R_1 = CH_3$ ; $R_2 = H$

$\gamma$ - tocotrienol: $R_1 = R_2 = H$

**Fig. (2).** Chemical structures of walnut tocopherols and tocotrienols.

Among walnut minerals, potassium is the major one [21]. The importance of adequate potassium intake (and reduced sodium intake) is well documented in the promotion of heartbeat and blood pressure control and consequently in the reduction of the risk of hypertension and stroke [22, 23]. Other walnut minerals include phosphorus, magnesium, calcium, sodium, zinc and iron [21]. The intake of potassium, magnesium and calcium is beneficial against bone demineralization protection, hypertension, insulin resistance, and cardiovascular complications [21].

## *Phytochemicals*

Walnuts are rich in melatonin [12], a natural antioxidant which seems to protect the seed from oxidative stress (OS) [24]. In animals, this hormone produced by the pineal gland, regulates the responses to circadian rhythms, promoting healthy sleep patterns, and also acts as a potent antiradical capacity against reactive oxygen species (ROS) and reactive nitrogen species (RNS), facilitating the immune system response and retarding the aging process [24, 25].

Walnuts are also a significant dietary source of polyphenols in the human diet (115 mg/g of methanolic extract) [3]. Epidemiological studies suggest that ingestion of natural products rich in antioxidants, such as flavonoids and phenolic acids, can prevent the development several diseases [9, 26, 27]. Phenolic acids have a wide spectrum of pharmacological activities, namely antiradicalar, antimutagenic, antitumoral and anticancer properties [3, 9, 28 - 33]. Antiviral and antibacterial properties of phenolic acids have also been observed in some studies [1, 34]. Additionally, it has been demonstrated that flavonoids possess antiinflammatory, hepatoprotector, antitumoral, antibacterial, antiviral and inhibition of enzymes capacities [26, 32, 35]. Many authors have devoted themselves to the analysis of the phenolic composition of several walnut cultivars and verified that the phenolics most commonly present are: several phenolic acids (Fig. **3**), such as gallic, ellagic, syringic, *p*-coumaric, ferulic, sinapic, caffeic and 5-*O*-caffeoylquinic acids, juglone (Fig. **4**), a naftoquinone, and tannins, namely stenophyllarin, glansrins A, B and C, and casuarinin [11, 36].

Walnut leaves phenolic compounds have also been studied and its total content is also considerable (95 mg/g of methanolic extract) [3]. Amaral *et al.* [18] and Pereira *et al.* [1] studied the phenolic composition of various cultivars and found various compounds, namely 3-*O*-caffeoylquinic, 5-*O*-caffeoylquinic, *p*-coumaric, 3-*O*-*p*-coumaroylquinic and 4-*O*-*p*-coumaroylquinic acids, quercetin 3-*O*-galactoside, quercetin 3-*O*-arabinoside, quercetin 3-*O*-xyloside, and quercetin 3-*O*-ramnoside.

Studies about WGH are still very scarce. However, its total phenolic content is also considerable (about 50 mg/g of methanolic extract) [3]. According to Stampar *et al.* [7], WGH phenolic profile is composed by thirteen compounds, such as chlorogenic acid, caffeic acid, ferulic acid, sinapic acid, gallic acid, ellagic acid, protocatechuic acid, syringic acid, vanillic acid, catechin, epicatechin, myricetin and juglone, which is the major phenolic compound.

BENZOIC ACIDS DERIVATIVES

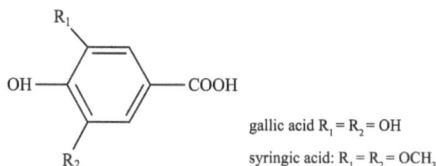

gallic acid $R_1 = R_2 = OH$
syringic acid: $R_1 = R_2 = OCH_3$

CINNAMIC ACID DERIVATIVES

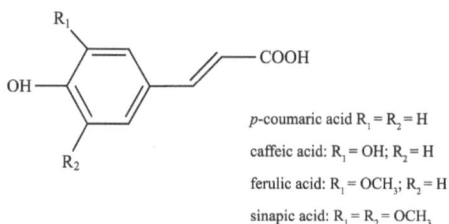

*p*-coumaric acid $R_1 = R_2 = H$
caffeic acid: $R_1 = OH; R_2 = H$
ferulic acid: $R_1 = OCH_3; R_2 = H$
sinapic acid: $R_1 = R_2 = OCH_3$

**Fig. (3).** Chemical structures of walnut phenolic acids.

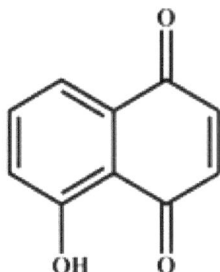

**Fig. (4).** Chemical structure of juglone.

## ANTIOXIDANT ACTIVITY AND ASSOCIATED HEALTH BENEFITS

OS has been associated to development and progression of wide range of diseases [37, 38]. Oxidation is a natural process occurring in most living organisms. Nevertheless, they present enzymatic and non-enzymatic defences against an excessive production of ROS. Glutathione and thioredoxin are examples of endougenous antioxidant defences.

The occurrence of mutations in genes coding these endogenous antioxidants can lead to increase of OS-related diseases and to premature death [9, 39, 40]. Furthermore, aging and some external factors, such as drugs, smoke, alcohol and even high-caloric foods can impair endogenous antioxidant defences and, consequently, alter redox equilibrium [32]. ROS in large amounts can cause DNA, lipids and proteins damages [41, 42]. OS happens when there is an imbalance between ROS production and the antioxidant defence capacity against them. This modifies the major components of cells and tissues [43]. Thus, since antioxidants are capable of scavenging ROS, they are crucial against development and progression of OS-mediated diseases [44].

Noteworthy, natural antioxidants can scavenge free radicals inhibiting oxidation [45, 46]. Interestingly, walnuts have a great antioxidant capacity, mainly caused by their high phenolic content, and thus, can play a crucial role in the protection against these deleterious diseases [30]. Phenolic compounds are effective antioxidants because of their chemical structure, namely due to phenolic hydroxyl groups hydrogen-donating capacity [47]. Espín *et al.* [48] studied the antiradicalar activity of walnut oil when compared with several other types of oils.

The authors related that high antiradicalar activity is linked with the high phenolic content it contains. Since then, walnut polyphenols have been found to inhibit human plasma and LDL oxidation [49]. Several studies have confirmed a strong association between total polyphenolic content and antioxidant potential [20, 50 - 52]. As previously discussed, walnuts possess one of the highest amount of total phenolic content [9] and thus one of the highest antioxidant activities [53] among all nut types. Walnut is also rich in ellagic acid, which has antiproliferative and antioxidant properties, as shown in studies *in vitro* and with animal models [54, 55]. Additionally, it also prevents the binding of DNA to some carcinogens [56, 57]. Ellagic acid, similarly to other polyphenol antioxidants, can reduce OS in cellular models, exerting a chemoprotective effect [54]. Very recently, Chen *et al.* [58] revealed for the first time that this acid is able to suppress cell cycle in the G0/G1 phase, inhibiting the proliferation of MCF-7 breast cancer cells.

*J. regia* also contains flavonoids. These phytochemicals are potent antioxidants and showed to reduce the ROS level in RAW264.7 cells [59]. Strong antioxidant activities have been reported on walnut seeds [36], leaves [1] and green husks [6]. Recently, Carvalho *et al.* [3] tested and compared the free radical scavenging capacity of two different types of extracts, methanolic and petroleum ether extracts, of walnut seeds, leaves and WGH.

The highest total phenolic content and 2,2'-diphenyl-1-picrylhydrazyl (DPPH) radical scavenging activity were found in walnuts methanolic extract, followed by leaves and WGH methanolic extracts. The antioxidant action was very reduced or absent in petroleum ether extracts. Methanolic extracts, specially leaves extract, protected the erythrocyte membrane from hemolysis induced by 2,2'-azobis (2-amidinopropane) dihydrochloride (AAPH). In this study, a relation between antioxidant power and total phenolic content was confirmed. By comparing the scavenging effect of methanolic and petroleum ether extracts, Carvalho *et al.* [3] concluded that the polyphenols in walnut derivatives have a more potent antioxidant activity than non-polar antioxidants of the lipid fraction, such as tocopherols and tocotrienols. Thus, polyphenols are probably the most important compounds in the antiradicalar protection conferred by walnut derivatives. The antioxidant properties of the phenolic compounds are due to their redox properties, acting as reducing agents, hydrogen donors, free radical scavengers, singlet oxygen quenchers, and metal chelators [60]. Due to these properties, these phytochemicals have been used to counter several diseases related to OS, namely diabetes, neurodegenerative and cardiovascular diseases and cancer [32, 37, 38]. Since it possesses a high content of phenolic compounds, *J. regia* presents itself as a chemopreventive agent against OS-related diseases. Several different types of *J. regia* extracts have been tested in diabetic models and patients, due to its high phenolic content and antioxidant activity. In the last decades, diabetes has been linked to ROS [61]. It is known that OS is relevent in the development and progression of this metabolic disease [62]. Diabetic individuals have an increased production of free radicals and impaired antioxidant defences [63]. ROS are produced by reduction of molecular oxygen, and oxidation of water producing superoxide anion, and hydroxyl radicals and hydrogen peroxide. In a recent study in type 2 diabetic patients, a *J. regia* leaves extract has shown promising effects, significantly decreasing fasting blood glucose levels, the percentage of glycated haemoglobin and total cholesterol of the diabetic patients without important adverse effects [64]. These effects were achieved with an ingestion of *J. regia* ethanolic leaf extract twice a day, during 3 months. In an effort to understand the exact mechanism of this antidiabetic effect, Pitschmann *et al.* [65] aimed ate evaluating the effect of a methanolic extract in one of the most promising drug targets for type 2 diabetes mellitus patients, protein–tyrosine phosphatase 1B (PTP1B). Inhibitors of this cytosolic enzyme not only increase cellular response to insulin, but also elevate leptin signalling and is therefore a promising therapeutic strategy for both diabetes and obesity [66]. However, the results obtained were contradictory, and more evidence is needed for more reliable conclusions.

## ANTIMUTAGENIC AND ANTICARCIONAGENIC ACTIVITIES

Cancer is one of the major health problems nowadays, affecting millions of people worldwide [67]. It results from an uncontrolled cell division that causes formation of tumours. The initiation stage happens when carcinogens injure the DNA. If the damage in the DNA is not repaired, this leads to genetic mutations. These initiated cells, then begin a deregulated expansion, tissue remodelling and inflammation, originating tumours [33]. When they reach this phase, tumours normally have several changes in cancer-related genes [68]. These effects are dependent of a number of factors, such as life styles, pollution or mutations [69].

Despite being a major cause of deaths nowadays, cancer can be largely preventable by healthy life styles, particularly in terms of dietary factors [33].

Data from the literature suggest that the most cancers are linked to mitogens and mutagens [70]. Because of this, it is important that chemoprevention strategies focus on substances that can act in different cellular processes [33]. Moreover, OS is present in several types of cancer cells, with OS-related DNA damage being involved in development of mutations, carcinogenesis, and ageing. DNA damage is an essential step in cancer initiation and progression. In fact, higher levels of oxidative damage of DNA have been reported in several types of tumours, wich strongly indicates that damage plays a key role in the aetiology of cancer [37].

Phytomedicine is a good and helpful research area for the fight against cancer. Medicinal plants are used in chemoprevention and chemotherapy without serious side effects [71 - 73]. As discussed, *J. regia* has a strong antioxidant effect due to its phytochemical composition, whether in its seeds, leaves or green husks. As such, and coupled with the high phenolic content, *J. regia* seems like a promising chemo-protective agent. In fact, several studies have already been performed to investigate the effectiveness of *J. regia* and its major compounds as anticancer agents. Salimi *et al.* [74] showed that 5,7-dihydroxy-3,4′-dimethoxyflavone and regiolone have anti-proliferative effects in human breast and oral cancer cell lines. Juglone is other compound that is known to lower blood glucose and has a potent anticancer effect by inhibiting tumor proliferation, mediating G2 phase cell cycle arrest and inducing apoptosis [53, 75 - 77]. Additionally, Polonik *et al.* [78] reported that juglone has potent immunostimulating activity in tumorous tissues. However, the anticancer effects of juglone are usually attributed to induction of OS, depletion of glutathione, and damaging of cell membranes, which causes cell death by apoptosis and necrosis [79 - 81]. Additionally, juglone inhibits peptidyl-prolyl isomerase Pin1, which is commonly overexpressed in many types of human cancer [82]. Other effects include the suppression of the activator protein-2α and the human epidermal growth factor receptor-2 promoter [83], the downregulation of p53 protein levels [84], the alteration of B-cell lymphoma 2 (Bcl-2) and Bax levels [79, 80, 85], the activation of PARP [79] and the activation of subsequent caspase [86, 87]. Moreover, juglone has been known to prevent cancer cells from entering mitosis for a long time, as juglone-treated tumours usually present sticky and diffused chromosomes in the prophase, and an accumulation of abnormal metaphase figures [88].

A juglone-rich diet has also been found to lower the incidence of azoxymethane induced intestinal tumors [89]. Moreover, juglone is reportedly an inhibitor of peptidyl-propyl isomerase Pin1, which depletion in cancer cells causes mitotic arrest and apoptosis [90, 91]. The overexpression of Pin1 has also been observed in various types of cancers [92], and has been linked to an increase in tumorigenesis caused by the Ras and Neu oncogenes, and to the regulation of the proliferative capacity [93]. Pin1 has been recently receiving attention as a potential therapeutic agent for cancer [94], since juglone is a strong inhibitor of it, being that *J. regia* can be of great use as a source of juglone. Noteworthy, the antiproliferative effect of juglone seems to be mediated by an increase in OS. However, *J. regia* extracts possess a high antioxidant activity in serveral assays, illustring that the effects of juglone is different when used as a single compound or as part of an extract. In fact, the synergetic effect with other phytocomponents can prove to be a significant ally in the fight against cancer.

Carvalho *et al.* [3] have evaluated the effect of methanolic extracts of walnut seeds, leaves and WGH in the inhibition of human renal and colon cancer cells proliferation. Notably, all different types of extracts displayed concentration-dependent growth inhibition in both types of cells. In A-498 cell line, all extracts presented similar antiproliferative activity, while leaves extract showed the highest antiproliferative efficiency for both 769-P renal and Caco-2 colon cancer cells. In addition, recently Alshatwi *et al.* [95] have reported the antiproliferative effects of WGH extracts on human prostate cancer cells *via* apoptosis induction. These results need to be interpreted taking in consideration the chemical composition of the walnut plant, which includes omega 3 fatty acids, vitamin E, phytosterols, ellagic acid, gallic acid and flavonoids, namely quercetin, carotenoids, and melatonin, all of which have been linked with antiproliferative effects in cancer cells [96 - 102].

Interestingly, the main contributors for the reported beneficial effects of walnut plant are thought to be polyphenols [51, 103, 104]. Polyphenols can act as anticancer agents by supressing the activation of transcription factors such as factor nuclear kappa B (NF-kB). NF-kB is responsible for regulation of genes involved in inflammation and carcinogenesis. Polyphenols also suppress the activation of transcription factor activator protein-1 (AP-1), a molecular target in chemoprevention and that has increased activity in several types of cancer. Additionally, polyphenols also suppress protein kinases (PK), such as PKC, and growth factor receptor mediated pathways. Moreover, the antioxidant, antiangiogenic and antiinflammatory properties of this type of compounds can also be involved [3, 33, 105]. However, and since no correlation has been found between total phenolic compound content and anti-proliferative activity, it is probable that the anticancer properties of *J. regia* are a result of additive or even synergistic interactions between the numerous compounds present in the plant extracts [3, 106].

## CONCLUDING REMARKS

In conclusion, walnut tree stands up as an excellent and economical source of cost-effective natural antioxidant phytochemicals with chemoprotective properties. However, despite all the beneficial effects of *J. regia* seeds, green husks and leafsreported in this work, there still is a major lack of knowledge of all the complex molecular mechanisms involved in the interaction between cancer cells and *J. regia* and its phytocomponents. Additional in-depth studies are required if we pretend to full comprehend how this interaction occurs. On that account, we expect that this work has contributed to the increase of studies about *J. regia*, promoting its ingestion. We believed that it can be a promising therapy against cancer development.

## CONFLICT OF INTEREST

The authors confirm that they have no conflict of interest to declare for this publication.

## ACKNOWLEDGEMENTS

This work was supported by the "Fundação para a Ciência e a Tecnologia"–FCT (PEst-C/SAU/UI0709/2014) co-funded by Fundo Europeu de Desenvolvimento Regional - FEDER *via* Programa Operacional Factores de Competitividade -

COMPETE/QREN. M.G. Alves (SFRH/BPD/80451/2011) was funded by FCT. P.F. Oliveira was funded by FCT through FSE and POPH funds (PTDC/QUI-BIQ/121446/2010).

## REFERENCES

[1]     Pereira JA, Oliveira I, Sousa A, *et al.* Walnut (Juglans regia L.) leaves: phenolic compounds, antibacterial activity and antioxidant potential of different cultivars. Food Chem Toxicol 2007; 45(11): 2287-95.
[http://dx.doi.org/10.1016/j.fct.2007.06.004] [PMID: 17637491]

[2]     Fernandéz J, Aleta N, Alía R. Forest genetic resources conservation of Juglans regia L. Noble hardwoods network report of the fourth meeting gmunden. In: International Plant Genetic Resources Institute; Italy. 1999.

[3]     Carvalho M, Ferreira PJ, Mendes VS, *et al.* Human cancer cell antiproliferative and antioxidant activities of Juglans regia L. Food Chem Toxicol 2010; 48(1): 441-7.
[http://dx.doi.org/10.1016/j.fct.2009.10.043] [PMID: 19883717]

[4]     Clark AM, Jurgens TM, Hufford CD. Antimicrobial activity of juglone. Phytother Res 1990; 4: 11-4.
[http://dx.doi.org/10.1002/ptr.2650040104]

[5]     Dreher ML, Maher CV, Kearney P. The traditional and emerging role of nuts in healthful diets. Nutr Rev 1996; 54(8): 241-5.
[http://dx.doi.org/10.1111/j.1753-4887.1996.tb03941.x] [PMID: 8961751]

[6]     Oliveira I, Sousa A, Ferreira IC, Bento A, Estevinho L, Pereira JA. Total phenols, antioxidant potential and antimicrobial activity of walnut (Juglans regia L.) green husks. Food Chem Toxicol 2008; 46(7): 2326-31.
[http://dx.doi.org/10.1016/j.fct.2008.03.017] [PMID: 18448225]

[7]     Stampar F, Solar A, Hudina M, Veberic R, Colaric M. Traditional walnut liqueur–cocktail of phenolics. Food Chem 2006; 95: 627-31.
[http://dx.doi.org/10.1016/j.foodchem.2005.01.035]

[8]     Amaral JS, Casal S, Pereira JA, Seabra RM, Oliveira BP. Determination of sterol and fatty acid compositions, oxidative stability, and nutritional value of six walnut (Juglans regia L.) cultivars grown in Portugal. J Agric Food Chem 2003; 51(26): 7698-702.
[http://dx.doi.org/10.1021/jf030451d] [PMID: 14664531]

[9]     Blomhoff R, Carlsen MH, Andersen LF, Jacobs DR Jr. Health benefits of nuts: potential role of antioxidants. Br J Nutr 2006; 96 (Suppl. 2): S52-60.
[http://dx.doi.org/10.1017/BJN20061864] [PMID: 17125534]

[10]    Martínez ML, Labuckas DO, Lamarque AL, Maestri DM. Walnut (Juglans regia L.): genetic resources, chemistry, by-products. J Sci Food Agric 2010; 90(12): 1959-67.
[PMID: 20586084]

[11]    Colaric M, Veberic R, Solar A, Hudina M, Stampar F. Phenolic acids, syringaldehyde, and juglone in fruits of different cultivars of Juglans regia L. J Agric Food Chem 2005; 53(16): 6390-6.
[http://dx.doi.org/10.1021/jf050721n] [PMID: 16076123]

[12]    Pereira JA, Oliveira I, Sousa A, Ferreira IC, Bento A, Estevinho L. Bioactive properties and chemical composition of six walnut (Juglans regia L.) cultivars. Food Chem Toxicol 2008; 46(6): 2103-11. [http://dx.doi.org/10.1016/j.fct.2008.02.002] [PMID: 18334279]

[13]    Van Hellemont J. Compendium de Phytotherapie, Association Pharmaceutique. Bruxelles, Belge: APB. Serv Sci 1986.

[14]    Bruneton J. Pharmacognosie, Phytochimie, Plantes Médicinales 4ª edition. Paris: Tec & Doc/Lavoisier 2009.

[15]    Wichtl M, Anton R. Plantes thérapeutiques 2ª edition. Paris: Tec & Doc 1999.

[16]    Valnet J. Phytothérapie: traitement des maladies par las plantes. Maloine 1983.

[17]    Girzu M, Carnat A, Privat A-M, Fialip J, Carnat AP, Lamaison JL. Sedative Effect of Walnut Leaf Extract and Juglone, an Isolated Constituent. Pharm Biol 1998; 36(4): 280-6. [http://dx.doi.org/10.1076/phbi.36.4.280.4580]

[18]    Amaral JS, Seabra RM, Andrade PB, Valentao P, Pereira JA, Ferreres F. Phenolic profile in the quality control of walnut (Juglans regia L.) leaves. Food Chem 2004; 88: 373-9. [http://dx.doi.org/10.1016/j.foodchem.2004.01.055]

[19]    Amaral JS, Alves MR, Seabra RM, Oliveira BP. Vitamin E composition of walnuts (Juglans regia L.): a 3-year comparative study of different cultivars. J Agric Food Chem 2005; 53(13): 5467-72. [http://dx.doi.org/10.1021/jf050342u] [PMID: 15969535]

[20]    Li L, Tsao R, Yang R, Kramer JK, Hernandez M. Fatty acid profiles, tocopherol contents, and antioxidant activities of heartnut (Juglans ailanthifolia Var. cordiformis) and Persian walnut (Juglans regia L.). J Agric Food Chem 2007; 55(4): 1164-9. [http://dx.doi.org/10.1021/jf062322d] [PMID: 17253708]

[21]    Ros E. Health benefits of nut consumption. Nutrients 2010; 2(7): 652-82. [http://dx.doi.org/10.3390/nu2070652] [PMID: 22254047]

[22]    McCune LM, Kubota C, Stendell-Hollis NR, Thomson CA. Cherries and health: a review. Crit Rev Food Sci Nutr 2011; 51(1): 1-12. [http://dx.doi.org/10.1080/10408390903001719] [PMID: 21229414]

[23]    Ferreira S, Zanella M, Freire M, Milagres R, Plavinik F, Ribeiro A. Blood pressure management in diabetic patients. Nefrologia 1994; 14.

[24]    Zhao Y, Tan DX, Lei Q, *et al.* Melatonin and its potential biological functions in the fruits of sweet cherry. J Pineal Res 2013; 55(1): 79-88. [http://dx.doi.org/10.1111/jpi.12044] [PMID: 23480341]

[25]    Rocha CS, Martins AD, Rato L, Silva BM, Oliveira PF, Alves MG. Melatonin alters the glycolytic profile of Sertoli cells: implications for male fertility. Mol Hum Reprod 2014; 20(11): 1067-76. [http://dx.doi.org/10.1093/molehr/gau080] [PMID: 25205674]

[26]    Nour V, Trandafir I, Cosmulescu S. HPLC determination of phenolic acids, flavonoids and juglone in walnut leaves. J Chromatogr Sci 2013; 51(9): 883-90. [http://dx.doi.org/10.1093/chromsci/bms180] [PMID: 23135132]

[27]    Davis PA, Vasu VT, Gohil K, *et al.* A high-fat diet containing whole walnuts (Juglans regia) reduces

tumour size and growth along with plasma insulin-like growth factor 1 in the transgenic adenocarcinoma of the mouse prostate model. Br J Nutr 2012; 108(10): 1764-72.
[http://dx.doi.org/10.1017/S0007114511007288] [PMID: 22244053]

[28]    Akbari V, Jamei R, Heidari R, Esfahlan AJ. Antiradical activity of different parts of Walnut (Juglans regia L.) fruit as a function of genotype. Food Chem 2012; 135(4): 2404-10.
[http://dx.doi.org/10.1016/j.foodchem.2012.07.030] [PMID: 22980820]

[29]    Alarcón E, Campos AM, Edwards AM, Lissi E, López-Alarcón C. Antioxidant capacity of herbal infusions and tea extracts: A comparison of ORAC-fluorescein and ORAC-pyrogallol red methodologies. Food Chem 2008; 107: 1114-9.
[http://dx.doi.org/10.1016/j.foodchem.2007.09.035]

[30]    Almajano MP, Carbo R, Jiménez J, Gordon MH. Antioxidant and antimicrobial activities of tea infusions. Food Chem 2008; 108: 55-63.
[http://dx.doi.org/10.1016/j.foodchem.2007.10.040]

[31]    Atoui AK, Mansouri A, Boskou G, Kefalas P. Tea and herbal infusions: Their antioxidant activity and phenolic profile. Food Chem 2005; 89: 27-36.
[http://dx.doi.org/10.1016/j.foodchem.2004.01.075]

[32]    Dias T, Tomás G, Teixeira N, Alves MG, Oliveira PF, Silva BM. White tea (Camellia Sinensis (L.)): antioxidant properties and beneficial health effects. Int J Food Sci Nutr Diet 2013; 2: 1-15.

[33]    Fresco P, Borges F, Diniz C, Marques MP. New insights on the anticancer properties of dietary polyphenols. Med Res Rev 2006; 26(6): 747-66.
[http://dx.doi.org/10.1002/med.20060] [PMID: 16710860]

[34]    Ziaková A, Brandšteterová E. Validation of HPLC determination of phenolic acids present in some Lamiaceae family plants. J Liquid Chromatogr Relat Technol 2003; 26: 443-53.
[http://dx.doi.org/10.1081/JLC-120017181]

[35]    Amaral JS, Valentão P, Andrade PB, Martins RC, Seabra RM. Do cultivar, geographical location and crop season influence phenolic profile of walnut leaves? Molecules 2008; 13(6): 1321-32.
[http://dx.doi.org/10.3390/molecules13061321] [PMID: 18596658]

[36]    Fukuda T, Ito H, Yoshida T. Antioxidative polyphenols from walnuts (Juglans regia L.). Phytochemistry 2003; 63(7): 795-801.
[http://dx.doi.org/10.1016/S0031-9422(03)00333-9] [PMID: 12877921]

[37]    Valko M, Leibfritz D, Moncol J, Cronin MT, Mazur M, Telser J. Free radicals and antioxidants in normal physiological functions and human disease. Int J Biochem Cell Biol 2007; 39(1): 44-84.
[http://dx.doi.org/10.1016/j.biocel.2006.07.001] [PMID: 16978905]

[38]    Valko M, Rhodes CJ, Moncol J, Izakovic M, Mazur M. Free radicals, metals and antioxidants in oxidative stress-induced cancer. Chem Biol Interact 2006; 160(1): 1-40.
[http://dx.doi.org/10.1016/j.cbi.2005.12.009] [PMID: 16430879]

[39]    Dalton TP, Chen Y, Schneider SN, Nebert DW, Shertzer HG. Genetically altered mice to evaluate glutathione homeostasis in health and disease. Free Radic Biol Med 2004; 37(10): 1511-26.
[http://dx.doi.org/10.1016/j.freeradbiomed.2004.06.040] [PMID: 15477003]

[40]    Valentine JS, Doucette PA, Zittin Potter S. Copper-zinc superoxide dismutase and amyotrophic lateral

sclerosis. Annu Rev Biochem 2005; 74: 563-93.
[http://dx.doi.org/10.1146/annurev.biochem.72.121801.161647] [PMID: 15952898]

[41]   Halliwell B. Antioxidants and human disease: a general introduction. Nutr Rev 1997; 55(1): S44-9.
[PMID: 9155225]

[42]   Sohal RS, Weindruch R. Oxidative stress, caloric restriction, and aging. Science 1996; 273(5271): 59-63.
[http://dx.doi.org/10.1126/science.273.5271.59] [PMID: 8658196]

[43]   Almeida IF, Fernandes E, Lima JL, Costa PC, Bahia MF. Walnut (Juglans regia) leaf extracts are strong scavengers of pro-oxidant reactive species. Food Chem 2008; 106: 1014-20.
[http://dx.doi.org/10.1016/j.foodchem.2007.07.017]

[44]   Willett WC. Diet and health: what should we eat? Science 1994; 264(5158): 532-7.
[http://dx.doi.org/10.1126/science.8160011] [PMID: 8160011]

[45]   Wongkham S, Laupattarakasaem P, Pienthaweechai K, Areejitranusorn P, Wongkham C, Techanitiswad T. Antimicrobial activity of Streblus asper leaf extract. Phytother Res 2001; 15(2): 119-21.
[http://dx.doi.org/10.1002/ptr.705] [PMID: 11268109]

[46]   Jin D, Hakamata H, Takahashi K, Kotani A, Kusu F. Determination of quercetin in human plasma after ingestion of commercial canned green tea by semi-micro HPLC with electrochemical detection. Biomed Chromatogr 2004; 18(9): 662-6.
[http://dx.doi.org/10.1002/bmc.370] [PMID: 15386501]

[47]   Lindsay DG, Astley SB. European research on the functional effects of dietary antioxidants - EUROFEDA. Mol Aspects Med 2002; 23(1-3): 1-38.
[http://dx.doi.org/10.1016/S0098-2997(02)00005-5] [PMID: 12079769]

[48]   Espín JC, Soler-Rivas C, Wichers HJ. Characterization of the total free radical scavenger capacity of vegetable oils and oil fractions using 2,2-diphenyl-1-picrylhydrazyl radical. J Agric Food Chem 2000; 48(3): 648-56.
[http://dx.doi.org/10.1021/jf9908188] [PMID: 10725129]

[49]   Anderson KJ, Teuber SS, Gobeille A, Cremin P, Waterhouse AL, Steinberg FM. Walnut polyphenolics inhibit *in vitro* human plasma and LDL oxidation. J Nutr 2001; 131(11): 2837-42.
[PMID: 11694605]

[50]   Li L, Tsao R, Yang R, Liu C, Zhu H, Young JC. Polyphenolic profiles and antioxidant activities of heartnut (Juglans ailanthifolia Var. cordiformis) and Persian walnut (Juglans regia L.). J Agric Food Chem 2006; 54(21): 8033-40.
[http://dx.doi.org/10.1021/jf0612171] [PMID: 17032006]

[51]   Silva BM, Andrade PB, Valentão P, Ferreres F, Seabra RM, Ferreira MA. Quince (Cydonia oblonga Miller) fruit (pulp, peel, and seed) and Jam: antioxidant activity. J Agric Food Chem 2004; 52(15): 4705-12.
[http://dx.doi.org/10.1021/jf040057v] [PMID: 15264903]

[52]   Parry J, Su L, Moore J, *et al.* Chemical compositions, antioxidant capacities, and antiproliferative activities of selected fruit seed flours. J Agric Food Chem 2006; 54(11): 3773-8.

[http://dx.doi.org/10.1021/jf060325k] [PMID: 16719495]

[53]　Bolling BW, McKay DL, Blumberg JB. The phytochemical composition and antioxidant actions of tree nuts. Asia Pac J Clin Nutr 2010; 19(1): 117-23.
[PMID: 20199996]

[54]　Seeram NP, Adams LS, Henning SM, *et al. In vitro* antiproliferative, apoptotic and antioxidant activities of punicalagin, ellagic acid and a total pomegranate tannin extract are enhanced in combination with other polyphenols as found in pomegranate juice. J Nutr Biochem 2005; 16(6): 360-7.
[http://dx.doi.org/10.1016/j.jnutbio.2005.01.006] [PMID: 15936648]

[55]　Narayanan BA, Geoffroy O, Willingham MC, Re GG, Nixon DW. p53/p21(WAF1/CIP1) expression and its possible role in G1 arrest and apoptosis in ellagic acid treated cancer cells. Cancer Lett 1999; 136(2): 215-21.
[http://dx.doi.org/10.1016/S0304-3835(98)00323-1] [PMID: 10355751]

[56]　Mandal S, Stoner GD. Inhibition of N-nitrosobenzylmethylamine-induced esophageal tumorigenesis in rats by ellagic acid. Carcinogenesis 1990; 11(1): 55-61.
[http://dx.doi.org/10.1093/carcin/11.1.55] [PMID: 2295128]

[57]　Teel RW, Babcock MS, Dixit R, Stoner GD. Ellagic acid toxicity and interaction with benzo[a]pyrene and benzo[a]pyrene 7,8-dihydrodiol in human bronchial epithelial cells. Cell Biol Toxicol 1986; 2(1): 53-62.
[http://dx.doi.org/10.1007/BF00117707] [PMID: 3267445]

[58]　Chen HS, Bai MH, Zhang T, Li G-D, Liu M. Ellagic acid induces cell cycle arrest and apoptosis through TGF-β/Smad3 signaling pathway in human breast cancer MCF-7 cells. Int J Oncol 2015; 46(4): 1730-8.
[PMID: 25647396]

[59]　Zhao MH, Jiang ZT. Flavonoids in *Juglans regia* L. leaves and evaluation of *in vitro* antioxidant activity *via* intracellular and chemical methods. Scient World J 2014; 2014: 303878.

[60]　Costa RM, Magalhães AS, Pereira JA, *et al.* Evaluation of free radical-scavenging and antihemolytic activities of quince (Cydonia oblonga) leaf: a comparative study with green tea (Camellia sinensis). Food Chem Toxicol 2009; 47(4): 860-5.
[http://dx.doi.org/10.1016/j.fct.2009.01.019] [PMID: 19271320]

[61]　Baynes JW. Role of oxidative stress in development of complications in diabetes. Diabetes 1991; 40(4): 405-12.
[http://dx.doi.org/10.2337/diab.40.4.405] [PMID: 2010041]

[62]　Ceriello A. Oxidative stress and glycemic regulation. Metabolism 2000; 49(2) (Suppl. 1): 27-9.
[http://dx.doi.org/10.1016/S0026-0495(00)80082-7] [PMID: 10693917]

[63]　Bloch-Damti A, Bashan N. Proposed mechanisms for the induction of insulin resistance by oxidative stress. Antioxid Redox Signal 2005; 7(11-12): 1553-67.
[http://dx.doi.org/10.1089/ars.2005.7.1553] [PMID: 16356119]

[64]　Hosseini S, Jamshidi L, Mehrzadi S, *et al.* Effects of Juglans regia L. leaf extract on hyperglycemia and lipid profiles in type two diabetic patients: a randomized double-blind, placebo-controlled clinical

trial. J Ethnopharmacol 2014; 152(3): 451-6.
[http://dx.doi.org/10.1016/j.jep.2014.01.012] [PMID: 24462785]

[65]    Pitschmann A, Zehl M, Atanasov AG, Dirsch VM, Heiss E, Glasl S. Walnut leaf extract inhibits PTP1B and enhances glucose-uptake *in vitro*. J Ethnopharmacol 2014; 152(3): 599-602.
[http://dx.doi.org/10.1016/j.jep.2014.02.017] [PMID: 24548753]

[66]    Thareja S, Aggarwal S, Bhardwaj TR, Kumar M. Protein tyrosine phosphatase 1B inhibitors: a molecular level legitimate approach for the management of diabetes mellitus. Med Res Rev 2012; 32(3): 459-517.
[http://dx.doi.org/10.1002/med.20219] [PMID: 20814956]

[67]    Siegel R, Naishadham D, Jemal A. Cancer statistics, 2012. CA Cancer J Clin 2012; 62(1): 10-29.
[http://dx.doi.org/10.3322/caac.20138] [PMID: 22237781]

[68]    Stanley LA. Molecular aspects of chemical carcinogenesis: the roles of oncogenes and tumour suppressor genes. Toxicology 1995; 96(3): 173-94.
[http://dx.doi.org/10.1016/0300-483X(94)02991-3] [PMID: 7900159]

[69]    Noonan DM, Benelli R, Albini A. Angiogenesis and cancer prevention: a vision. Recent Results Cancer Res 2007; 174: 219-24.
[http://dx.doi.org/10.1007/978-3-540-37696-5_19] [PMID: 17302199]

[70]    Kelloff GJ, Hawk ET, Karp JE, *et al.* Progress in clinical chemoprevention. Semin Oncol 1997; 24(2): 241-52.
[PMID: 9129692]

[71]    Conde VR, Alves MG, Oliveira PF, Silva BM. Tea (Camellia Sinensis (L.)): a Putative Anticancer Agent in Bladder Carcinoma? Anticancer Agents Med Chem 2014; 15: 26-36.
[http://dx.doi.org/10.2174/1566524014666141203143143] [PMID: 25495463]

[72]    Carvalho M, Jerónimo C, Valentão P, Andrade PB, Silva BM. Green tea: A promising anticancer agent for renal cell carcinoma. Food Chem 2010; 122: 49-54.
[http://dx.doi.org/10.1016/j.foodchem.2010.02.014]

[73]    Moderno PM, Carvalho M, Silva BM. Recent patents on Camellia sinensis: source of health promoting compounds. Recent Pat Food Nutr Agric 2009; 1(3): 182-92.
[http://dx.doi.org/10.2174/2212798410901030182] [PMID: 20653539]

[74]    Salimi M, Ardestaniyan MH, Mostafapour Kandelous H, *et al.* Anti-proliferative and apoptotic activities of constituents of chloroform extract of Juglans regia leaves. Cell Prolif 2014; 47(2): 172-9.
[http://dx.doi.org/10.1111/cpr.12090] [PMID: 24467376]

[75]    Inbaraj JJ, Chignell CF. Cytotoxic action of juglone and plumbagin: a mechanistic study using HaCaT keratinocytes. Chem Res Toxicol 2004; 17(1): 55-62.
[http://dx.doi.org/10.1021/tx034132s] [PMID: 14727919]

[76]    Kim SH, Lee KS, Son JK, *et al.* Cytotoxic compounds from the roots of Juglans mandshurica. J Nat Prod 1998; 61(5): 643-5.
[http://dx.doi.org/10.1021/np970413m] [PMID: 9599266]

[77]    Kamei H, Koide T, Kojima T, Hashimoto Y, Hasegawa M. Inhibition of cell growth in culture by quinones. Cancer Biother Radiopharm 1998; 13(3): 185-8.

[http://dx.doi.org/10.1089/cbr.1998.13.185] [PMID: 10850354]

[78]    Polonik S, Prokof'eva N, Agafonova I, Uvarova N. Antitumor and immunostimulating activity of 5-hydroxy-1, 4-naphthoquinone (juglone) O-and S-acetylglycosides. Pharm Chem J 2003; 37: 397-8.
[http://dx.doi.org/10.1023/A:1027305110622]

[79]    Seshadri P, Rajaram A, Rajaram R. Plumbagin and juglone induce caspase-3-dependent apoptosis involving the mitochondria through ROS generation in human peripheral blood lymphocytes. Free Radic Biol Med 2011; 51(11): 2090-107.
[http://dx.doi.org/10.1016/j.freeradbiomed.2011.09.009] [PMID: 21982843]

[80]    Ji Y-B, Qu Z-Y, Zou X. Juglone-induced apoptosis in human gastric cancer SGC-7901 cells *via* the mitochondrial pathway. Exp Toxicol Pathol 2011; 63(1-2): 69-78.
[http://dx.doi.org/10.1016/j.etp.2009.09.010] [PMID: 19815401]

[81]    Aithal BK, Kumar MR, Rao BN, Udupa N, Rao BS. Juglone, a naphthoquinone from walnut, exerts cytotoxic and genotoxic effects against cultured melanoma tumor cells. Cell Biol Int 2009; 33(10): 1039-49.
[http://dx.doi.org/10.1016/j.cellbi.2009.06.018] [PMID: 19555768]

[82]    Fila C, Metz C, van der Sluijs P. Juglone inactivates cysteine-rich proteins required for progression through mitosis. J Biol Chem 2008; 283(31): 21714-24.
[http://dx.doi.org/10.1074/jbc.M710264200] [PMID: 18539601]

[83]    Khanal P, Namgoong GM, Kang BS, Woo ER, Choi HS. The prolyl isomerase Pin1 enhances HER-2 expression and cellular transformation *via* its interaction with mitogen-activated protein kinase/extracellular signal-regulated kinase kinase 1. Mol Cancer Ther 2010; 9(3): 606-16.
[http://dx.doi.org/10.1158/1535-7163.MCT-09-0560] [PMID: 20179161]

[84]    Paulsen MT, Ljungman M. The natural toxin juglone causes degradation of p53 and induces rapid H2AX phosphorylation and cell death in human fibroblasts. Toxicol Appl Pharmacol 2005; 209(1): 1-9.
[http://dx.doi.org/10.1016/j.taap.2005.03.005] [PMID: 16271620]

[85]    Xu HL, Yu XF, Qu SC, *et al.* Anti-proliferative effect of Juglone from Juglans mandshurica Maxim on human leukemia cell HL-60 by inducing apoptosis through the mitochondria-dependent pathway. Eur J Pharmacol 2010; 645(1-3): 14-22.
[http://dx.doi.org/10.1016/j.ejphar.2010.06.072] [PMID: 20655907]

[86]    Montenegro RC, Araújo AJ, Molina MT, *et al.* Cytotoxic activity of naphthoquinones with special emphasis on juglone and its 5-O-methyl derivative. Chem Biol Interact 2010; 184(3): 439-48.
[http://dx.doi.org/10.1016/j.cbi.2010.01.041] [PMID: 20138029]

[87]    Lu JJ, Bao JL, Wu GS, *et al.* Quinones derived from plant secondary metabolites as anti-cancer agents. Anticancer Agents Med Chem 2013; 13(3): 456-63.
[PMID: 22931417]

[88]    Okada TA, Roberts E, Brodie AF. Mitotic Abnormalities Produced by Juglone in Ehrlich Ascites Tumor Cells. Proceedings of the Society for Experimental Biology and Medicine Society for Experimental Biology and Medicine. New York. 1967.
[http://dx.doi.org/10.3181/00379727-126-32513]

[89]     Sugie S, Okamoto K, Rahman KM, *et al.* Inhibitory effects of plumbagin and juglone on azoxymethane-induced intestinal carcinogenesis in rats. Cancer Lett 1998; 127(1-2): 177-83.
         [http://dx.doi.org/10.1016/S0304-3835(98)00035-4] [PMID: 9619875]

[90]     Chao S-H, Greenleaf AL, Price DH. Juglone, an inhibitor of the peptidyl-prolyl isomerase Pin1, also directly blocks transcription. Nucleic Acids Res 2001; 29(3): 767-73.
         [http://dx.doi.org/10.1093/nar/29.3.767] [PMID: 11160900]

[91]     Zhang CJ, Zhang ZH, Xu BL, Wang YL. [Recent advances in the study of pin1 and its inhibitors]. Yao Xue Xue Bao 2008; 43(1): 9-17.
         [PMID: 18357725]

[92]     Thakur A. A therapeutic phytochemical from Juglans regia L. J Med Plants Res 2011; 5: 5324-30.

[93]     Finn G, Lu KP. Phosphorylation-specific prolyl isomerase Pin1 as a new diagnostic and therapeutic target for cancer. Curr Cancer Drug Targets 2008; 8(3): 223-9.
         [http://dx.doi.org/10.2174/156800908784293622] [PMID: 18473735]

[94]     Mathur R, Chandna S, N. Kapoor P, S. Dwarakanath B. Peptidy prolyl isomerase, Pin1 is a potential target for enhancing the therapeutic efficacy of etoposide. Curr Cancer Drug Targets 2011; 11(3): 380-90.
         [http://dx.doi.org/http://dx.doi.org/10.2174/156800911794519761] [PMID: 21247380]

[95]     Alshatwi AA, Hasan TN, Shafi G, *et al.* Validation of the Antiproliferative Effects of Organic Extracts from the Green Husk of Juglans regia L. on PC-3 Human Prostate Cancer Cells by Assessment of Apoptosis-Related Genes. Evid Based Complement Altern Med 2012; 2012: 103026.

[96]     Berquin IM, Edwards IJ, Chen YQ. Multi-targeted therapy of cancer by omega-3 fatty acids. Cancer Lett 2008; 269(2): 363-77.
         [http://dx.doi.org/10.1016/j.canlet.2008.03.044] [PMID: 18479809]

[97]     Bradford PG, Awad AB. Phytosterols as anticancer compounds. Mol Nutr Food Res 2007; 51(2): 161-70.
         [http://dx.doi.org/10.1002/mnfr.200600164] [PMID: 17266177]

[98]     Spaccarotella KJ, Kris-Etherton PM, Stone WL, *et al.* The effect of walnut intake on factors related to prostate and vascular health in older men. Nutr J 2008; 7: 13.
         [http://dx.doi.org/10.1186/1475-2891-7-13] [PMID: 18454862]

[99]     Han DH, Lee MJ, Kim JH. Antioxidant and apoptosis-inducing activities of ellagic acid. Anticancer Res 2006; 26(5A): 3601-6.
         [PMID: 17094489]

[100]   Yáñez J, Vicente V, Alcaraz M, *et al.* Cytotoxicity and antiproliferative activities of several phenolic compounds against three melanocytes cell lines: relationship between structure and activity. Nutr Cancer 2004; 49(2): 191-9.
         [http://dx.doi.org/10.1207/s15327914nc4902_11] [PMID: 15489212]

[101]   Bhuvaneswari V, Nagini S. Lycopene: a review of its potential as an anticancer agent. Curr Med Chem Anticancer Agents 2005; 5(6): 627-35.
         [http://dx.doi.org/10.2174/156801105774574667] [PMID: 16305484]

[102]  Srinivasan V, Spence DW, Pandi-Perumal SR, Trakht I, Cardinali DP. Therapeutic actions of melatonin in cancer: possible mechanisms. Integr Cancer Ther 2008; 7(3): 189-203.
[http://dx.doi.org/10.1177/1534735408322846] [PMID: 18815150]

[103]  Fiorentino A, D'Abrosca B, Pacifico S, *et al.* Isolation and structure elucidation of antioxidant polyphenols from quince (Cydonia vulgaris) peels. J Agric Food Chem 2008; 56(8): 2660-7.
[http://dx.doi.org/10.1021/jf800059r] [PMID: 18348529]

[104]  Russo GL. Ins and outs of dietary phytochemicals in cancer chemoprevention. Biochem Pharmacol 2007; 74(4): 533-44.
[http://dx.doi.org/10.1016/j.bcp.2007.02.014] [PMID: 17382300]

[105]  Bonfili L, Cecarini V, Amici M, *et al.* Natural polyphenols as proteasome modulators and their role as anti-cancer compounds. FEBS J 2008; 275(22): 5512-26.
[http://dx.doi.org/10.1111/j.1742-4658.2008.06696.x] [PMID: 18959740]

[106]  Mertens-Talcott SU, Bomser JA, Romero C, Talcott ST, Percival SS. Ellagic acid potentiates the effect of quercetin on p21waf1/cip1, p53, and MAP-kinases without affecting intracellular generation of reactive oxygen species *in vitro*. J Nutr 2005; 135(3): 609-14.
[PMID: 15735102]

# CHAPTER 6

# Bioactive Compounds of Chestnuts as Health Promoters

**Teresa Delgado[1,2], José A. Pereira[1], Susana Casal[2,\*], Elsa Ramalhosa[1,\*]**

[1] *Mountain Research Centre (CIMO) - School of Agriculture, Polytechnic Institute of Bragança, Bragança, Portugal*

[2] *LAQV-REQUIMTE, Chemistry Department, Faculty of Pharmacy, Oporto University, Porto, Portugal*

**Abstract:** Different chestnut species can be cultivated for fruit production, the most valorised part for nutritional purposes. However *Castanea sativa* Mill., the "European chestnut", is one of the most valorised worldwide. Its fruits are consumed either raw or after processing, being boiling and roasting the most usual ones. The nutritional composition of fresh chestnut is variable, with interesting amounts of carbohydrates and fibre, together with low fat content, with differences between cultivars and producing regions. In respect to the presence of bioactive compounds, such as phenolic compounds, vitamins, fatty acids, among others, some studies had focused on the fruit benefits to human health but few reported the effect of processing in those compounds. In this context, this chapter intended to review the current knowledge on chestnut composition, together with the influence of diverse post-harvest technologies, such as refrigeration, flame peeling, freezing with $CO_2$, irradiation, boiling and roasting on the bioactive compounds of chestnut.

**Keywords:** Antioxidant activity, Bioactive compounds, Boiling, Carbohydrates, *Castanea sativa* Miller, Cold storage, Drying, Fatty acids, Fibre, Irradiation, Minerals, Nutritional composition, Organic acids, Osmotic dehydration, Phenolic compounds, Processing, Proteins, Roasting, Vitamin C, Vitamin E.

\* **Corresponding authors Elsa Ramalhosa and Susana Casal:** Mountain Research Centre (CIMO) - School of Agriculture, Polytechnic Institute of Bragança, Apartado 1172, 5301-85 Bragança, Portugal; Tel: +351 273303308; Fax: +351 273325405; E-mail:elsa@ipb.pt and LAQV-REQUIMTE, Chemistry Department, Faculty of Pharmacy, Oporto University, Porto, Portugal; E-mail:sucasal@ff.up.pt.

## INTRODUCTION

The genus *Castanea* belongs to the angiosperm family *Fagaceae*. Throughout the world, several different species of chestnut can be found, such as *Castanea creanata* Sieb. in Asia, *C. creanata* Zucc. in Japan, *C. mollissima* Bl. in China and Korea, *C. dentada* Borkh in North America, and *C. sativa* Mill. in Europe being, also called "European chestnut" [1, 2].

The specie *C. sativa* Mill. is one of the most valorised worldwide. However, to improve chestnut production and the resistance to certain common diseases of the tree, some hybrids have emerged over the years [3].

Chestnut production has a high importance in the world's primary economy. China is the main producer, with about 1650000 tonnes (t) in 2012, followed by Republic of Korea (70000 t), Turkey (59789 t), Boli*via* (57000 t), Italy (52000 t), Greece (28700 t), Japan (20900 t) and Portugal, the eighth largest world's producer (19100 t) [4]. Despite these figures, due to the small country size, chestnut production in Portugal still represents a high contribution for the trade balance. The greatest production area for this fruit is located in Trás-os-Montes region (northeast of Portugal). Being a natural product, chestnut production can be affected by several factors including climatic conditions such as temperature, sunlight and precipitation, and also cultivation inputs, for exemple nutrients, minerals, and diseases and pests [3].

From a botanical point of view, chestnut fruit is a starchy nut composed by a seed protected by a membrane called the pellicle (episperm) and followed by a brown peel called "shell". This last involves the nut that is shielded by a spiny bur. When the fruit begins to mature, the bur modifies its colour, from green to yellow-brown, and breaks in 2-4 lengthways lines liberating three nuts. Sometimes the bur releases the chestnut fruits from the tree; however, more often the bur falls and opens completely on the ground as a result of the high humidity, liberating the fruits [5] (Fig. **1**). Even though the shell and pellicle are difficult to remove, this nut presents interesting properties which will be presented throughout this chapter.

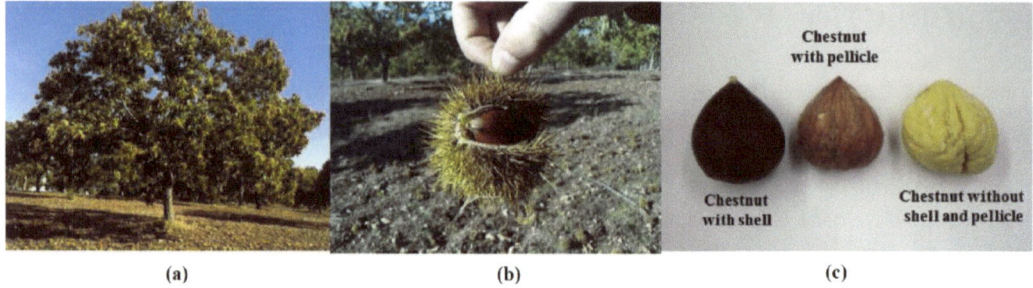

Fig. (1). *Castanea sativa*: **(a)** Tree, **(b)** Chestnut fruit in the bur, **(c)** Fruit.

There are some review studies that discuss the nutritional properties of chestnuts *in natura* [3, 6]; however, despite not being usual to consume it raw, there are few works reporting the effects of different types of processing on its physicochemical properties and bioactivity. Thus, in this chapter we intend to evidence the beneficial effects of chestnut on human health and in which way processing may affect the physicochemical properties of this fruit.

## PHYSICOCHEMICAL PROPERTIES OF CHESTNUT

Numerous studies on the physicochemical characterization of different varieties of chestnuts from different countries have been published. From the nutritional point of view, chestnut has interesting properties. Chestnuts are a good source of fibres, starch, protein, aminoacids, minerals, lipids, vitamin E and phenolic compounds, being a naturally gluten-free product.

The proximate nutritional composition of raw chestnuts(*Castanea sativa* Mill.) is detailed in Table **1**. The major compound is water, with moisture ranging between 40 and 64 g/100 g fresh weight (FW). This high moisture content represents a strong disadvantage for long-term preservation purposes, due to the high probability of mould formation, together with a significant weight loss during storage. On dry basis (DM), carbohydrates are the main components of chestnuts (75-91%), particularly starch (39 and 82%). Several studies have been reported about the specific content of amylose and amylopectin [15, 19, 20], accounting approximately for 33% and 67% of the starch content, respectively. The starch can improve health by giving energy from the catabolism process of amylose and amylopectin into glucose, as well as it can have a positive role on gut functions due to the presence of short-chain fatty acids (SCFA) that result from the bacterial catabolism of amylopectin-derived dextrins [19, 21]. Also, some researchers defend that starch is partially hydrolysed into glucose during storage, which gives increased sweetness to chestnuts [10, 15]. Indeed, this is one of the most important quality parameters, with recognized differences between varieties [22], being the sweetest fruits highly appreciated by the consumers. Sucrose is the main free sugar in chestnuts (Table **2**), followed by glucose and fructose. Other sugars such as maltose and xylose can also be present in chestnut fruits, but in reduced amounts.

**Table 1. Proximate composition of raw chestnut fruits from several cultivars and origins (*Castanea sativa* Mill).**

| Moisture (g/100g FW) | Carbohydrates (g/100g DM) | Starch (g/100g DM) | Total minerals (g/100g DM) | Crude protein (g/100g DM) | Crude fat (g/100g DM) | Energetic value (kcal/100g DM) | References |
|---|---|---|---|---|---|---|---|
| 40-64 | - | 39-67 | 1.5-8.0 | 3.9-10.9 | 0.9-4.4 | - | [3] |
| 52-55 | - | - | 1.5-1.9 | 5.0-6.5 | 1.6-1.8 | 401-402 | [7] |
| 53-54 | 89-91 | - | 1.7-2.4 | 5.2-6.4 | 1.5-1.7 | 400-402 | [8] |
| 46-53 | - | 39-48 | 1.5-2.2 | 4.9-7.4 | 1.7-3.1 | - | [9] |
| - | 75-86 | 54-70 | 1.0-3.2 | 4.9-10.9 | 0.5-2.0 | - | [10] |
| 46-51 | - | 64-65 | 1.9-2.3 | 3.9-5.1 | 1.6-1.9 | - | [11] |
| 41-54 | - | 49-54 | 2.0-2.7 | - | 1.9-4.4 | 413-427 | [12] |
| - | - | - | - | 4.1-5.8 | - | - | [13] |
| 48-59 | - | 57-82 | 1.8-3.1 | 6.0-8.6 | 1.3-3.0 | - | [14] |
| 40-60 | - | 42-66 | 1.8-3.2 | 4.5-9.6 | 1.7-4.0 | - | [15] |
| 48-57 | - | 49-52 | 2.1-2.4 | 4.2-7.0 | 2.2-4.5 | - | [16] |
| 50-60 | - | - | 1.7-2.6 | 4.5-9.9 | 1.9-4.7 | - | [17] |
| 42-53 | - | - | 1.2-2.4 | 4.2-5.6 | 3.0-4.6 | - | [18] |
| 40-64 | 75-91 | 39-82 | 1.0-8.0 | 3.9-10.9 | 0.5-4.7 | 400-427 | Min-Max |

| Fibre (g/100g DM) | | | | References |
|---|---|---|---|---|
| NDF | ADF | ADL | Celulose | |
| 2.7-28.9 | 0.5-4.5 | 0.02-1.3 | 0.5-3.6 | [3] |
| 2.7-3.8 | 0.5-0.6 | 0.02 | 0.5-0.6 | [7] |
| 3.2-3.7 | 0.5-0.6 | - | 0.5 | [8] |
| 13.8-24.4 | 1.9-3.2 | - | - | [9] |
| 13.1-13.8 | 2.5-2.7 | 0.2-0.3 | 2.3-2.4 | [11] |
| 13.8-19.9 | 3.0-3.6 | 0.4-0.6 | 2.6-3.1 | [12] |
| 9.4-28.5 | 2.3-4.5 | | | [15] |
| 2.7-28.9 | 0.5-4.5 | 0.02-1.3 | 0.5-3.6 | Min-Max |

Fibre is also an important nutrient in chestnuts (Table **1**). Among fibre fractions neutral detergent fibre (NDF) represents 2.7 to 28.9% and acid detergent fibre (ADF) varies from 0.5 to 4.5%, while acid detergent lignin (ADL) only represents 0.02 to 1.3% and cellulose 0.5 to 3.6%, all on dry matter (DM) basis. Dietary fibre has been described as the fragments of plant components which are not hydrolysed by human enzymes and it includes cellulose, hemicellulose and lignin, as well as other substances, namely waxes, cutin, and suberin [25].

**Table 2. Free sugar contents of raw chestnut fruits (*Castanea sativa* Mill.) (g/100g DM).**

| Sucrose | Glucose | Fructose | Maltose | Xylose | References |
|---|---|---|---|---|---|
| 6.6-29.7 | <DL -2.3 | 0.04-2.3 | 0-1.8 | | [3] |
| 8.9-21.3 | - | - | - | | [10] |
| 6.6-19.5 | <DL -0.3 | 0.04-0.3 | | | [14] |
| 5.1-9.8 | 0.1-0.3 | 0.05-0.2 | - | - | [16] |
| 10.9-22.1 | | 0.44-2.22 | | | [18] |
| 29.7 | 1.4 | 1.9 | 1.5 | 0.4 | [20] |
| 3.7-24.2 | 1.0-6.8 | 0.6-5.3 | - | | [22] |
| 6.6-19.5 | <DL -0.3 | 0.04-0.3 | - | | [23] |
| 10.2-13.7 | 0.07-0.3 | 0.11-0.4 | 0.4-0.6 | | [24] |
| 3.7-29.7 | 0.07-6.8 | 0.04-5.3 | 0-1.8 | 0.4 | Min-Max |

DL –Detection limit

Dietary fibres are associated to positive health effects, including "stimulation of *Bifidobacterium* and *Lactobacillus* in the intestine, decrease in cholesterol levels, reduction of the risk of cardiovascular diseases, positive regulation of insulin response, increase in anticancer mechanisms and positive effects on metabolism of blood lipids" [26]. When the fermentation of dietary fibre occurs SCFA are produced. These compounds are very important to garantee the maintenance of colonic integrity and metabolism. They can also be considered therapeutic agents for the treatment of some illnesses such as colitis, antibiotic associated diarrhoea and colon cancer [27].

Few studies have been performed on proteins in chestnuts; however, proteins range from 3.9 to 10.9 g/100 g DM (Table **1**), being identified some essential amino acids such as arginine (Arg), isoleucine (Ile), leucine (Leu), phenylalanine (Phe), threonine (Thr), tryptophan (Trp) and valine (Val), and non-essential amino acids such as alanine (Ala), asparagine (Asn), aspartic acid (Asp), glutamine (Gln), glutamic acid (Glu), glycine (Gly), serine (Ser) and tyrosine (Tyr) [11, 13].

Concerning minerals (as total ashes), these vary from 1.0 to 8.0 g/100 g DM, being detected important macro-elements (K, P, Mg, Ca, S and Na). The most important in this group is potassium. Some interesting micro-elements (Fe, Mn,

Zn, Cu, B and Se) have also been identified (Table **3**). Mineral intake has an important role in human health. Calcium has important biological functions, namely giving rigidity to the skeleton [3]. Magnesium has an important enzymatic role (cofactor), is involved in the synthesis of some compounds (namely proteins, RNA and DNA), and keeps the electrical potential of nervous tissues and cell membranes [3]. Micro-elements also play important roles in health. For example, iron is related to the transport of oxygen through red blood cell haemoglobin. Moreover, iron and zinc are responsible by the synthesis and catabolism of some nutrients and xenobiotics, once are essential components of enzymes that participate in these processes [28].

**Table 3. Mineral contents of raw chestnut fruits (*Castanea sativa* Mill.) (g/100g DM).**

| K | P | Mg | Ca | S | Na | Fe | Mn | Zn | Cu | B | Se | References |
|---|---|---|---|---|---|---|---|---|---|---|---|---|
| 473-1476 | 68-305 | 47-100 | 26-72 | 26-133 | 0.8-30.9 | 1.4-10.9 | 1.5-12.5 | 0.6-3.1 | 0.4-2.7 | 3.0-3.1 | 0.4-0.8 | [3] |
| 473-974 | 104-148 | 63-93 | 41-51 | . | 0.8-3.9 | 5.3-10.9 | 3.1-8.0 | 1.4-3.1 | 1.3-2.7 | . | . | [9] |
| 761-1271 | 107-191 | 70-160 | 43-230 | . | 6-41 | 0.4-5.7 | 0.7-5.5 | 1.8-9.1 | 0.6-3.8 | . | . | [10] |
| 789-1130 | 68-305 | 49-100 | 26-72 | . | 3.0-26 | 1.4-2.4 | 1.7-12.5 | 1.0-1.9 | 0.6-1.0 | . | . | [15] |
| 740-940 | 106-159 | 54-66 | 55-70 | . | . | 2.1-6.1 | 1.3-4.1 | 0.9-1.7 | 0.3-2.3 | . | . | [16] |
| 633-811 | 110-142 | 49-58 | 31-46 | 35-46 | 21.4-28.3 | 3.0-7.1 | 1.5-5.1 | 0.8-1.1 | 0.6-0.9 | 3.0-3.1 | 0.5-0.7 | [24] |
| 473-1476 | 68-305 | 47-160 | 26-230 | 26-133 | 0.8-41 | 0.4-10.9 | 0.7-12.5 | 0.6-9.1 | 0.3-3.8 | 3.0-3.1 | 0.4-0.8 | Min-Max |

Some differences in protein and mineral contents can be found between varieties. These differences can be due to genetic differences, altitude and soil type, among others, as suggested by Pereira-Lorenzo *et al*. [15]. Míguelez *et al*. [14] refer that soils with a higher content of schist originate fruits with higher protein content than granite-based ones.

Although chestnuts present a low crude fat content (0.5-4.7 g/100 g DM) (Table **1**), its lipids are constituted by low saturated fatty acids (SFA) (14.1-27.7%) and high unsaturated fatty acids (USFA) contents, namely monounsaturated (MUFA) (17.9-40.8%) and polyunsaturated (PUFA) (41.5-60.1%) fatty acids. The main individual fatty acids in chestnuts are linoleic (C18:2) (37.6-51.4%), followed by oleic (C18:1) (17.4-38.2%), palmitic (C16:0) (12.0-17.3%), and α-linolenic acid (C18:3) (4-10.3%) (Table **4**). This fatty acids profile, low in saturated fatty acids and high in polyunsaturated ones, with the presence of omega 3 fatty acids, plays an important role in several physiological processes by regulating plasma lipid levels, neuronal development, and by having cardiovascular, immune and visual functions, as well as insulin action [31].

**Table 4. Fatty acid composition (g/100g fatty acids) of raw chestnut fruits (*Castanea sativa* Mill.).**

| SFA | C16:0 | MUFA | C18:1 | PUFA | C18:2 | C18:3 | References |
|---|---|---|---|---|---|---|---|
| 14.1-27.7 | - | 17.9-39.3 | 17.4-37.6 | 42.0-60.1 | 37.6-50.9 | - | [3] |
| 16.2-19.4 | 14.2-17.3 | 30.9-38.7 | 29.6-37.4 | 42.0-51.9 | 37.9-45.5 | 4.0-6.4 | [7] |
| 16-19 | 16-17 | 36-38 | 35-37 | 43-48 | 38-40 | 4-5 | [8] |
| 14.1-18.6 | 12.5-16.8 | 22.5-39.3 | 20.7-37.6 | 42.0-60.1 | 37.6-50.9 | 4.4-10.0 | [29] |
| 15.9-24.6 | 12.0-16.8 | 21.7-40.8 | 21.3-38.2 | 41.5-56.7 | 37.9-51.4 | 3.56-10.3 | [30] |
| 14.1-27.7 | 12.0-17.3 | 17.9-40.8 | 17.4-38.2 | 41.5-60.1 | 37.6-51.4 | 4-10.3 | Min-Max |

Chestnut fruits also contain vitamins, such as, vitamin E and C (Table **5**). Vitamin E has an effect to protect the unsaturated fatty acids from oxidation, being $\gamma$-tocopherol the major vitamer (0.38-2.73 mg/100g FW). From the nutritional point of view, vitamin E has shown several benefits to human health, minimizing the harmful effects of inflammatory diseases (*e.g.* rheumatoid arthritis or hepatitis) [35], fortifying the immune system and decreasing the risk of cancer [36], as well as a probable contribution to decrease the viral load in HIV-infected patients [37], and aid in the treatment of Parkinson's syndrome [38].

**Table 5. Vitamin C and E contents of raw chestnut fruits (*Castanea sativa* Mill.).**

| Vitamin E (mg/100 g FW) | | | | | | References |
|---|---|---|---|---|---|---|
| Tocopherol | | | Tocotrienols | | | |
| $\alpha$ | $\gamma$ | $\delta$ | $\gamma$ | $\delta$ | | |
| $2\times10^{-3}$-$1\times10^{-2}$ | 0.38-2.73 | $2\times10^{-2}$-$10\times10^{-2}$ | $14\times10^{-2}$-$39\times10^{-2}$ | $1\times10^{-3}$-$4\times10^{-3}$ | | [3] |
| $23\times10^{-4}$-$100\times10^{-4}$ | 0.38-0.46 | $216\times10^{-4}$-$285\times10^{-4}$ | $197\times10^{-4}$-$399\times10^{-4}$ | $14\times10^{-4}$-$32\times10^{-4}$ | | [8] |
| - | 0.41-2.30 | $2\times10^{-2}$-$10\times10^{-2}$ | - | - | | [24] |
| $22\times10^{-4}$-$100\times10^{-4}$ | 0.38-0.478 | $195\times10^{-4}$-$332\times10^{-4}$ | $141\times10^{-4}$-$418\times10^{-4}$ | $11\times10^{-4}$-$41\times10^{-4}$ | | [32] |
| $22\times10^{-4}$-$1\times10^{-2}$ | 0.38-2.73 | $195\times10^{-4}$-$10\times10^{-2}$ | $141\times10^{-4}$-$39\times10^{-2}$ | $11\times10^{-4}$-$41\times10^{-4}$ | | Min-Max |
| Vitamin C (mg/100 g DM) | | | | | | |
| Ascorbic Acid | | Dehydroascorbic acid | | | | References |
| | 30.8-40.2[a,b] | | | | | [3] |
| | 0.77-1.64[a] | | | | | [3] |
| 4.2-7.2 | | 3.1-6.8 | | | | [3] |
| 28-128 | | | | | | [18] |
| 4.7-6.7[b] | | 4.04-6.13[b] | | | | [24] |
| | 40.0-69.3[a] | | | | | [33] |
| 4.52-16.4 | | - | | | | [34] |

[a]Ascorbic + dehydroascorbic acids.
[b]Values are presented as mg 100 g$^{-1}$ FW.

Vitamin C is a term that is used for all compounds with biological activity of *L*-ascorbic acid. *L*-Ascorbic and *L*-dehydroascorbic acids are the main sources of vitamin C [39]. Both species are absorbed in the gastrointestinal tract [40]. Vitamin C is probably the most important hydrophilic antioxidant and it is believed to be of high importance for defence against diseases and degenerative processes produced by oxidative stress [41].

Besides ascorbic acid, other organic acids have been identified in chestnut fruits such as citric, malic, quinic, fumaric and oxalic acids (Table **6**). It was observed that the ranges of some organic acids were high probably due to the different extraction methods used. Organic acids might have a protective effect against multiple diseases as a result of their antioxidant activity [43]. Moreover, they can affect the organoleptic characteristics of fruits and vegetables such as the flavour [44].

**Table 6. Organic acids contents of raw chestnut fruits (*Castanea sativa* Mill.) (mg/g DM).**

| Citric acid | Malic acid | Quinic acid | Fumaric acid | Oxalic acid | Oxalic acid + *cis*-aconitic acid | Malic acid + quinic acid | References |
|---|---|---|---|---|---|---|---|
| 1.5-8.8 | 1.5-5.4 | - | - | - | - | - | [17] |
| - | 1.5-3.3 | - | - | - | - | - | [18] |
| 0.04-0.11 | - | - | $0.2 \times 10^{-3}$-$1.4 \times 10^{-3}$ | - | $1.3 \times 10^{-3}$-$7.1 \times 10^{-3}$ | $3.6 \times 10^{-2}$-$11 \times 10^{-2}$ | [34] |
| 12 | 5 | 13 | 0.4 | 0.7 | - | - | [42][a] |
| 0.04-12 | 1.5-5.4 | 13 | $0.2 \times 10^{-3}$-0.4 | 0.7 | $1.3 \times 10^{-3}$-$7.1 \times 10^{-3}$ | $3.6 \times 10^{-2}$-$11 \times 10^{-2}$ | Min-Max |

[a]The values refer to chestnuts irradiated at 0, 0.5, 1, 3 and 6 kGy.

The antioxidants present in natural products are associated to the prevention of certain human diseases caused by oxidative stress such as inflammatory diseases, ischemic diseases, cancer, hemochromatosis, emphysema, gastric ulcers, hypertension and preeclampsia, neurological diseases, alcoholism, smoking-related diseases and others [45]. Several studies have been performed on the antioxidant activity of different parts of chestnuts, namely, seed, leaf, flower, bur, outer and inner skin, and fruit. As can be observed from Table **7** different methods have been applied that difficult data comparison. Nevertheless the methods applied have included the scavenging activity on ABTS and DPPH radicals (measured through the decrease in ABTS and DPPH radical absorption after exposure to radical scavengers), reducing power (measured by the conversion of a $Fe^{3+}$/ferricyanide complex to the ferrous form), β-carotene bleaching inhibition (by neutralizing the linoleate-free radical and other free radicals formed in the system that attack the highly unsaturated β-carotene model), FRAP (measured by the capacity of reducing the Fe(III)/tripyridyltriazine complex to the ferrous form (blue colour) with an increase in absorbance at 593 nm), hemolysis inhibition (measured by the inhibition of erythrocyte hemolysis), hydroxyl radical scavenging activity (measured by inhibition of the hydroxyl radical generated by the Fenton reaction ($Fe^{2+}/H_2O_2$)), and TBARS inhibition (measured by the inhibition of the lipid peroxidation through the decrease of thiobarbituric acid reactive substances (TBARS)). For chestnuts, some studies reported that the antioxidant activity can vary between regions with different edaphoclimatic conditions [54, 55], being the coldest places those with the highest antioxidant activity. Indeed, severe climatic conditions might signalling for the plant defence mechanisms, including the production of important antioxidants, particularly phenolic compounds [54, 55]. Barreira *et al.* [47] reported that leaves, skins (outer and inner skins) and flowers presented the highest values of antioxidant activity comparing with chestnut fruits. When comparing chestnuts with other nuts it can be observed that the antioxidant activity of chestnut fruits [47] was of the same order of magnitude to those obtained for almond, peanut and pine nut [56, 57], while it was lower than those observed for hazelnut and walnut for the DPPH radical scavenging method. Nevertheless, the extraction methods were sometimes different. Similar results were also obtained for the reducing power method.

Due to their importance, the total phenol contents and some specific phenolic compounds determined in different chestnut parts, such as bur, flowers, leaves, outer shell, inner shell, and fruit are also presented in Table **8**. Gallic and ellagic acids are the main phenolic acids in fresh chestnut fruits; however, many other phenolic compounds have also been identified, such as vescalagin (0.06 to 0.10 mg/g FW), castalagin (0.41 to 0.82 mg/g FW), tannin T1 (0.06 to 0.09 mg/g FW), tannin T2 (0.05 to 0.09 mg/g FW), acutissimin A (0.05 to 0.08 mg/g FW) acutissimin B (0.41 to 0.51 mg/g FW) and ellagic acid derivatives (0.11 to 0.16 mg/g FW) in the outer shell of chestnuts [59]. In the inner shell of chestnuts these compounds are also detected at the following concentrations: vescalagin, 0.04 to 0.07 mg/g FW; castalagin, 0.07 to 0.21 mg/g FW; tannin T1, 0.03 to 0.07 mg/g FW; tannin T2, 0.03 to 0.04 mg/g FW; acutissimin A, 0.03 to 0.04 mg/g FW; acutissimin B, 0.04 to 0.09 mg/g FW; and ellagic acid derivatives, 0.0 to 0.01

mg/g fresh weight have also been detected [59]. Also Otles and Selek [55] have identified some phenolic compounds on chestnut fruits, namely, syringic+caffeic acids (0.002 to 0.02 mg/g FW), vanillic acid (0.15 to 0.92 mg/g FW), rutin (0.005 to 0.026 mg/g FW), catechin (0.024 to 0.13 mg/g FW), chlorogenic acid (0.004 to 0.12 mg/g Fw), *p*-coumaric acid (0.004 to 0.033 mg/g FW), ferulic acid (0.004 to 0.015 mg/g FW) and naringin (0.007 to 0.021 mg/g FW). These compounds have several positive effects on health, for example antioxidant properties, decrease of the risk of cardiovascular diseases, anticancer mechanisms and anti-inflammatory properties [60, 61]. The content of these compounds may vary due to several factors, such as climacteric conditions, soil type, precipitation and altitude. Dinis *et al.* [54] reported that the coldest ecotypes presented higher gallic and ellagic acid contents than the hottest ones for Judia cultivar, as already described for the antioxidant activity. Regarding flowers, Sapkota *et al.* [49] reported differences on phenolic content when comparing pre-bloom and full-bloom, showing higher phenolic content the flowers at pre-bloom than full-bloom.

**Table 7. Antioxidant activity of different parts of chestnuts.**

| Tissue | ABTS radical | DPPH radical | Reducing power | β-Carotene bleaching inhibition | FRAP | Hemolysis inhibition | Hydroxyl radical | TBARS inhibition | Unity | Reference |
|---|---|---|---|---|---|---|---|---|---|---|
| Leaf | - | 12.6-23.0 | - | - | - | - | - | - | μg mL (EC$_{50}$) | [46] |
| Leaf | - | 170 | 313 | 145 | - | 169 | - | 31.4 | μg mL (EC$_{50}$) | [47] |
| Leaf | - | 21.4 | - | - | - | - | 0 | - | % (0.2 mg extract mL solution) | [48] |
| Flower | - | 74.9 | 87.3 | 161 | - | 196 | - | 9.93 | μg mL (EC$_{50}$) | [47] |
| Flower | - | 45.14-119.36 | 0.494-0.772$^a$ | - | - | - | - | - | μg mL (EC$_{50}$) | [49] |
| Catkin | - | 38 | - | - | - | - | 43.6 | - | % (0.2 mg extract mL solution) | [48] |
| Bur | 1.33-3.80 | 0.92-3.42 | - | - | 0.708-2.261$^b$ | - | - | - | mmol TRE g extract | [50] |
| Outer skin | - | 39.7 | 55.1 | 133 | - | 91.4 | - | 7.87 | μg mL (EC$_{50}$) | [47] |
| Outer skin | - | 21.4 | - | - | - | - | 21.8 | - | % (0.2 mg extract mL solution) | [48] |
| Inner skin | - | 32.7 | 68.7 | 164 | - | 47.5 | - | 11.5 | μg mL (EC$_{50}$) | [47] |
| Shell | - | - | - | - | 47.5-380.8 | - | - | - | mmol AAE mg extract | [51] |
| Fruit | 4.77-8.15 | - | - | - | - | - | - | - | μmoles TRE g DM | [18] |
| Fruit | - | >10,000 | 9044 | 3632 | - | 3486 | - | 1117 | μg mL (EC$_{50}$) | [47] |
| Fruit | - | 0 | - | - | - | - | 5.5 | - | % (0.2 mg extract mL solution) | [48] |
| Fruit | 0.564-1.046 | - | - | - | - | - | - | - | mmol TRE kg | [52] |
| Fruit | - | 25.12-38.72 | 2.81-7.05 | 6.00-6.38 | - | - | - | 5.21-10.63 | mg mL (EC$_{50}$) | [53] |
| Fruit | 5.2-14.1 | 7.3-33.5 | - | - | 6.6-14.6 | 0.63-1.31 | - | - | mg g (EC$_{50}$) | [54] |
| Fruit | - | - | - | - | 9.08-14.15 | - | - | - | mM FeSO$_4$ g DM | [55] |

TRE - Trolox equivalent; AAE - Ascrobic acid equivalent;
$^a$Absorbance at 700 nm of 1 mg/mL solution; $^b$Value expressed in mmol AAE/g extract;

**Table 8. Total phenols and individual phenolic compound contents present in different parts of chestnuts.**

| Tissue | Total phenols | Unity | Gallic acid | Ellagic acid | Unity | References |
|---|---|---|---|---|---|---|
| Leaves | 103 | mg CE g$_{extract}$ | - | - | - | [47] |
| Flowers | 298 | mg CE g$_{extract}$ | - | - | - | [47] |
| Flowers | 251.6-467.9 | mg GAE g$_{extract}$ | - | - | - | [49] |
| Bur | 168.8-359.8 | mg GAE g$_{extract}$ | - | - | - | [50] |
| Bark | - | | - | 0.71-21.6 | mg g DM | [58] |
| Shell | 266-597 | mg GAE g$_{extract}$ | - | - | - | [51] |
| Outer shell | 510 | mg CE g$_{extract}$ | - | - | - | [47] |
| Outer shell | 61.9-84.9 | mg GAE g FW | 0.14-0.33 | 0.14-0.18 | mg g FW | [59] |
| Inner shell | 475 | mg CE g$_{extract}$ | - | - | - | [47] |
| Inner shell | 76.0-106.0 | mg GAE g FW | 0.22-0.34 | 0.03-0.07 | mg g FW | [59] |
| Pericarp | - | | - | 0.04-0.19 | mg g DM | [58] |
| Pellicle | - | | - | 0.03-0.091 | mg g DM | [58] |
| Fruit | 15.80-22.69 | mg GAE g FW | 3.46-9.07 | 2.71-9.64 | mg g FW | [11] |
| Fruit | 7.66-18.30 | mg GAE g FW | 8.03-24.89 | 7.28-47.78 | mg g FW | [13] |
| Fruit | 13.6-18.8 | mg GAE g DM | 0.00376-0.0204 | nd – 0.0249 | mg g DM | [17] |
| Fruit | 0.0872-0.157 | mg GAE g DM | - | - | - | [18] |
| Fruit | 3.73 | mg CE g$_{extract}$ | - | - | - | [47] |
| Fruit | 3.61-3.63 | mg GAE g$_{extract}$ | - | - | - | [53] |
| Fruit | 9.6-19.4 | mg GAE g DM | 4.1-29.0 | 6.2-11.9 | mg g DM | [54] |
| Fruit | 5.00-32.82 | mg GAE g DM | 0.0859-0.277 | 0.0116-0.0487 | mg g DM | [55] |
| Fruit | - | | - | tr-0.05 | mg g DM | [58] |

## PRINCIPAL POSTHARVEST TECHNOLOGIES

### Cold Storage

Chestnut fruits are a seasonal product and several post-harvest techniques have been applied to preserve them, being cold storage one of the most commonly used. However, some studies reported modifications on some physicochemical properties of the fruits along cold storage. An increase of the dry matter content due to water loss and NDF fibre content has been observed when comparing chestnuts subjected to cold storage during three months with fresh chestnuts [12]. On the other hand, a decrease on starch content was observed, explained by the enzymatic catabolism of starch into soluble sugars. Some studies reported that sucrose content can increase along storage time [24, 62 - 64], while glucose and fructose remained almost constant during storage [24, 62]. Nevertheless, Chenlo *et al.* [64] reported that fructose depended on the moisture content because at low moisture contents fructose practically disappeared while it increased in samples stored at high moisture contents.

Regarding to vitamin E, a significant increase on δ-tocopherol and γ-tocopherol contents in fresh weight was observed along cold storage [24]. On contrary, a decrease was verified on ascorbic acid concentration, while the dehydroascorbic acid content was not affected by the storage period. Also, an increase on the total phenols, gallic and ellagic acid contents was observed [13], but a decrease on the antioxidant activity determined by different methods was perceived after storage, with the exception of the β-carotene bleaching inhibition method for chestnut skins [65].

## Drying

Hot air drying is another common technology applied to preserve chestnut fruits. Along the years, several studies using different dehydration methods have been performed, being hot air drying the most widely used. The aim of these studies was to evaluate several technological aspects, including drying kinetics [66, 67], drying characteristics and energy requirement for dehydration [68], the effect of drying temperatures on morphological, chemical, thermal and rheological properties of chestnut flours [69 - 71], and the effect of drying followed by rehydration on different chestnut properties [20, 72, 73].

Some of these studies reported a degradation on chestnut colour (browning) originated by the drying process, being more pronounced at higher temperatures [66, 72]. This can be a disadvantage because this characteristic is one of the most appreciated by the consumers of this nut. Moreover, Attanasio *et al.* [20] reported that sucrose content decreased significantly after drying at 60 °C, due to its thermal degradation. They also reported modifications on morphological characteristics after drying and rehydration as the products seemed more shapeless and the open pore volume of starch granules on dried samples increased. Correia *et al.* [69] observed that drying temperature affected chestnut flour properties, also depending on chestnut variety.

Regarding to the effect of hot air drying on other nutrients and bioactive compounds, no studies have been performed until now.

## Osmotic Dehydration

Another dehydration method is osmotic dehydration. Along the years, some studies have been performed using different osmotic agents such as sodium chloride, glycerol, glucose and sucrose [73 - 79], with the aim to evaluate mass transfer, dehydration kinetics of chestnut fruits and the effect of this treatment on chestnuts physical properties such as colour, size, shape, weight reduction (WR), water loss (WL), solid gain (SG), normalized moisture content (NMC). Again, no studies on the effect of osmotic dehydration on nutrients and bioactive compounds have been performed until now.

## Irradiation

As previously mentioned, mould development is a particular concern in chestnut preservation due to its high moisture and carbohydrate contents. Chemical fumigation is one of the most effective disinfestation method but, due to the toxicity of the gases used for the operators and associated environmental problems, fumigation was banned in EU since 2010 [80, 81]. Thus, irradiation appears as an alternative technique, being nowadays considered a more environmentally friendly technology, meeting the food safety requirements [6]. Different types of irradiation have been tested along the years, being the most common the gamma irradiation and the electron beam irradiation.

Some studies on the effect of gamma irradiation on chestnut properties referred the existence of no differences on the nutritional composition [63], or on sugars and fatty acid compositions [82] of irradiated chestnuts. Moreover, irradiation may protect antioxidants such as tocopherols and phenolic compounds, as the antioxidant activity seemed to increase when compared with non-irradiated samples [65]. On the other hand, Carocho *et al.* [53] reported that irradiation dose had a significant role on the antioxidant activity, being 3kGy the dose that gave the best results with the highest phenolic content and the lowest $EC_{50}$ values for DPPH scavenging activity, β-carotene bleaching and TBARS inhibition, suggesting a higher antioxidant activity.

Regarding to electron beam irradiation, Carocho *et al.* [53] reported that the irradiated samples presented also higher phenolic content and antioxidant activity than non-irradiated ones, except for flavonoids whose content decreased. In this study the irradiation dose that gave the best results was 1 kGy. However, Carocho *et al.* [53] reported that the effect of electron beam irradiation on chemical and nutritional properties of chestnuts was very low.

## Industrial Processing and Other Ways to Prepare Chestnuts for Consumption

Few studies have been performed with the aim to evaluate the effect of industrial processing on chestnut properties, such as flame peeling and freezing with $CO_2$. Regarding the proximate composition, small variations were observed when comparing fresh chestnuts with those submitted to industrial processing. These industrial processing methods had some positive effects, such as an increase on crude energy and fibre (NDF, ADF and cellulose) contents [12], as well as in free sugar contents [24]. On the other hand, a negative effect on starch and to a lesser extent in fat content [12] was observed with a decrease on these compounds. Concerning the bioactive compounds, industrial processing contributed positively by promoting a significant increase on total phenols, gallic and ellagic acids contents [13], and also on tocopherols amounts, while some negative effects, namely a decrease on some free amino acids [13] and vitamin C contents [24], were observed.

Another well-known chestnut processing is roasting. Some studies have been performed on the roasting effect on chestnuts physicochemical properties. Künsch *et al.* [83] observed that this process had a little effect on the chemical composition of different chestnut varieties, since the amount of starch, sucrose and fatty acids remained the same. Gonçalves *et al.* [17] observed that roasted chestnuts presented higher protein, fibre, citric acid, gallic acid and total phenolic contents than raw chestnuts. Regarding the effect of roasting on the bioactive compounds few studies have been performed until now. Barros *et al.* [52] observed that some cultivars were more affected by roasting than others concerning the hydrophilic antioxidant activity.

Boiling is also another cooking process very used by chestnut consumers. As expected, boiling promoted a decrease in dry matter content, as well as a decrease in chestnut colour such as brightness [84]. When comparing raw with cooked chestnuts (roasted and boiled) some studies reported differences between them.

Gonçalves *et al.* [17] observed that boiled chestnuts showed higher fat, soluble fibre, gallic and ellagic acids and total phenolic contents than raw chestnuts. Nevertheless, the last ones had significantly higher malic acid content than cooked nuts. Ribeiro *et al.* [34] observed a decrease on vitamin C (ascorbic acid) content indicating that this vitamin was degraded during the boiling process due to the high temperatures. Barros *et al.* [33] also detected a decrease in vitamin C content. Although vitamin C losses between 25 to 45% and 2 to 77% were observed for boiling and roasting processes, respectively, chestnuts subjected to both processes may be still a good source of this vitamin once it may represent 16.2% to 22.4% and 19.4% to 26.8% of the recommended dietary intake for adults, respectively.

## CONCLUDING REMARKS

In conclusion, most of the studies until now performed had focused on physicochemical characteristics of raw chestnuts, being analysed different parts of this nut, namely flowers, leaves, outer and inner skins, bur, among others. The results point out that chestnut is a nut with important constituents such as antioxidants and vitamins, as well as with low fat and high starch contents, being also a gluten-free nut. However, generally this fruit is not consumed raw, being subjected to different types of processing, namely, roasting and boiling. There are some works that had studied the effect of processing on physicochemical properties, being some modifications observed. Nevertheless more studies should be done with the purpose to valorise and develop new products based on chestnuts.

## CONFLICT OF INTEREST

The authors confirm that they have no conflict of interest to declare for this publication.

## ACKNOWLEDGEMENTS

Teresa Delgado acknowledges the Fundação para a Ciência e Tecnologia (FCT) for the financial support through the PhD grant—SFRH/BD/82285/2011, CIMO through the Project PEst-OE/AGR/UI0690/2014, REQUIMTE through the Project "PEst-OE/AGR/UI0690/2014" to "UID/QUI/50006/2013" and "Project NORTE-

07-0124-FEDER-000069", as well as the POCTEP – Programa de cooperação Transfronteiriça Espanha – Portugal through the Project RED/AGROTEC – Experimentation network and transfer for development of agricultural and agro industrial sectors between Spain and Portugal.

## REFERENCES

[1]     Bounous G. The chestnut: a multipurpose resource for the new millennium. In: Abreu CG, Rosa E, Monteiro AA, Eds. Proceedings of the 3ʳᵈ International Chestnut Congress. Chaves, Portugal . 2005; pp. 33-40.
        [http://dx.doi.org/10.17660/ActaHortic.2005.693.1]

[2]     Lang P, Dane F, Kubisiak TL. Phylogeny of *Castanea* (*Fagaceae*) based on chloroplast *trn*T-L-F sequence data. Tree Genet Genomes 2006; 2: 132-9.
        [http://dx.doi.org/10.1007/s11295-006-0036-2]

[3]     De Vasconcelos MCBM, Bennett RN, Rosa EAS, Ferreira-Cardoso JV. Composition of European chestnut (*Castanea sativa* Mill.) and association with health effects: fresh and processed products. J Sci Food Agric 2010; 90(10): 1578-89.
        [http://dx.doi.org/10.1002/jsfa.4016] [PMID: 20564434]

[4]     FAO. Available from: http://faostat.fao.org , [acessed 2 Feb 2015];2015

[5]     Mencarelli F. Postharvest handling and storage of chestnuts Working document of the project: TCP/CPR/8925 "Integrated Pest Management and Storage of Chestnuts in XinXian county, Henan, China". ed., China: FAO 2001.

[6]     Antonio AL, Carocho M, Bento A, Quintana B, Botelho ML, Ferreira ICFR. Effects of gamma radiation on the biological, physico-chemical, nutritional and antioxidant parameters of chestnuts - a review. Food Chem Toxicol 2012; 50(9): 3234-42.
        [http://dx.doi.org/10.1016/j.fct.2012.06.024] [PMID: 22735498]

[7]     Barreira JCM, Casal S, Ferreira ICFR, Oliveira MBPP, Pereira JA. Nutritional, fatty acid and triacylglycerol profiles of *Castanea sativa* Mill. cultivars: a compositional and chemometric approach. J Agric Food Chem 2009; 57(7): 2836-42.
        [http://dx.doi.org/10.1021/jf803754u] [PMID: 19334758]

[8]     Barreira JCM, Casal S, Ferreira ICFR, Peres AM, Pereira JA, Oliveira MBPP. Chemical characterization of chestnut cultivars from three consecutive years: chemometrics and contribution for authentication. Food Chem Toxicol 2012; 50(7): 2311-7.
        [http://dx.doi.org/10.1016/j.fct.2012.04.008] [PMID: 22525865]

[9]     Borges O, Gonçalves B, Soeiro de Carvalho JL, Correia P, Silva AP. Nutritional quality of chestnut (*Castanea sativa* Mill.) cultivars from Portugal. Food Chem 2008; 106: 976-84.
        [http://dx.doi.org/10.1016/j.foodchem.2007.07.011]

[10]    Ertürk Ü, Mert C, Soylu A. Chemical composition of fruits of some important chestnut cultivars. Braz Arch Biol Technol 2006; 49: 183-8.
        [http://dx.doi.org/10.1590/S1516-89132006000300001]

[11]   De Vasconcelos MCBM, Bennett RN, Rosa EAS, Ferreira-Cardoso JV. Primary and secondary metabolite composition of kernels from three cultivars of Portuguese chestnut (*Castanea sativa* Mill.) at different stages of industrial transformation. J Agric Food Chem 2007; 55(9): 3508-16.
[http://dx.doi.org/10.1021/jf0629080] [PMID: 17407304]

[12]   De Vasconcelos MCBM, Bennett RN, Rosa EAS, Ferreira-Cardoso JV. Industrial processing effects on chestnut fruits (*Castanea sativa* Mill.). 1. Starch, fat, energy and fibre. Int J Food Sci Technol 2009; 44: 2606-12.
[http://dx.doi.org/10.1111/j.1365-2621.2009.02091.x]

[13]   De Vasconcelos MCBM, Bennett RN, Rosa EAS, Ferreira-Cardoso JV. Industrial processing effects on chestnut fruits (*Castanea sativa* Mill.). 2. Crude protein, free amino acids and phenolic phytochemicals. Int J Food Sci Technol 2009; 44: 2613-9.
[http://dx.doi.org/10.1111/j.1365-2621.2009.02092.x]

[14]   Míguelez JDLM, Bernárdez MM, Queijeiro JMG. Composition of varieties of chestnuts from Galicia (Spain). Food Chem 2004; 84: 401-4.
[http://dx.doi.org/10.1016/S0308-8146(03)00249-8]

[15]   Pereira-Lorenzo S, Ramos-Cabrer AM, Díaz-Hernández MB, Ciordia-Ara M, Ríos-Mesa D. Chemical composition of chestnut cultivars from Spain. Sci Hortic (Amsterdam) 2006; 107: 306-14.
[http://dx.doi.org/10.1016/j.scienta.2005.08.008]

[16]   Dinis L-T, Peixoto F, Ferreira-Cardoso JV, *et al.* Influence of the growing degree-days on chemical and technological properties of chestnut fruits (var."Judia"). CyTA-J Food 2012; 10: 216-24.
[http://dx.doi.org/10.1080/19476337.2011.631713]

[17]   Gonçalves B, Borges O, Costa HS, Bennett R, Santos M, Silva AP. Metabolite composition of Chestnut (*Castanea sativa* Mill.) upon cooking: Proximate analysis, fibre, organic acid and phenolics. Food Chem 2010; 122: 154-60.
[http://dx.doi.org/10.1016/j.foodchem.2010.02.032]

[18]   Neri L, Dimitri G, Sacchetti G. Chemical composition and antioxidant activity of cured chestnuts from three sweet chestnut (*Castanea sativa* Mill.) ecotypes from Italy. J Food Compos Anal 2010; 23: 23-9.
[http://dx.doi.org/10.1016/j.jfca.2009.03.002]

[19]   Pizzoferrato L, Rotilio G, Paci M. Modification of structure and digestibility of chestnut starch upon cooking: a solid state $^{13}$C CP MAS NMR and enzymatic degradation study. J Agric Food Chem 1999; 47(10): 4060-3.
[http://dx.doi.org/10.1021/jf9813182] [PMID: 10552765]

[20]   Attanasio G, Cinquanta L, Albanese D, Matteo MD. Effects of drying temperatures on physico-chemical properties of dried and rehydrated chestnuts (*Castanea sativa*). Food Chem 2004; 88: 583-90.
[http://dx.doi.org/10.1016/j.foodchem.2004.01.071]

[21]   Nichols BL, Avery S, Sen P, Swallow DM, Hahn D, Sterchi E. The maltase-glucoamylase gene: common ancestry to sucrase-isomaltase with complementary starch digestion activities. Proc Natl Acad Sci USA 2003; 100(3): 1432-7.
[http://dx.doi.org/10.1073/pnas.0237170100] [PMID: 12547908]

[22]   Barreira JCM, Pereira JA, Oliveira MBPP, Ferreira ICFR. Sugars profiles of different chestnut (*Castanea sativa* Mill.) and almond (*Prunus dulcis*) cultivars by HPLC-RI. Plant Foods Hum Nutr 2010; 65(1): 38-43.
       [http://dx.doi.org/10.1007/s11130-009-0147-7] [PMID: 20033298]

[23]   Bernárdez MM, Miguélez JDLM, Queijeiro JG. HPLC determination of sugars in varieties of chestnut fruits from Galicia (Spain). J Food Compos Anal 2004; 17: 63-7.
       [http://dx.doi.org/10.1016/S0889-1575(03)00093-0]

[24]   De Vasconcelos MCBM, Nunes F, Viguera CG, Bennett RN, Rosa EAS, Ferreira-Cardoso JV. Industrial processing effects on chestnut fruits (*Castanea sativa* Mill.). 3. Minerals, free sugars, carotenoids and antioxidant vitamins. Int J Food Sci Technol 2010; 45: 496-05.
       [http://dx.doi.org/10.1111/j.1365-2621.2009.02155.x]

[25]   DeVries JW, Prosky L, Li B, Cho S. A historical perspective on defining dietary fiber. Cereal Foods World 1999; 44: 367-9.

[26]   Prosky L. When is dietary fiber considered a functional food? Biofactors 2000; 12(1-4): 289-97.
       [http://dx.doi.org/10.1002/biof.5520120143] [PMID: 11216498]

[27]   Cook SI, Sellin JH. Review article: short chain fatty acids in health and disease. Aliment Pharmacol Ther 1998; 12(6): 499-507.
       [http://dx.doi.org/10.1046/j.1365-2036.1998.00337.x] [PMID: 9678808]

[28]   Vitamin and Mineral Requirements in Human Nutrition. 2nd ed., Chaina: FAO/WHO 2004. Available at: http://apps.who.int/iris/bitstream/10665/42716/1/9241546123.pdf.

[29]   Borges OP, Carvalho JS, Correia PR, Silva AP. Lipid and fatty acid profiles of *Castanea sativa* Mill. chestnuts of 17 native Portuguese cultivars. J Food Compos Anal 2007; 20: 80-9.
       [http://dx.doi.org/10.1016/j.jfca.2006.07.008]

[30]   España MSA, Galdón BR, Romero CD, Rodríguez ER. Fatty acid profile in varieties of chestnut fruits from Tenerife (Spain). CyTA-J Food 2011; 9: 77-81.
       [http://dx.doi.org/10.1080/19476331003686858]

[31]   Benatti P, Peluso G, Nicolai R, Calvani M. Polyunsaturated fatty acids: biochemical, nutritional and epigenetic properties. J Am Coll Nutr 2004; 23(4): 281-302.
       [http://dx.doi.org/10.1080/07315724.2004.10719371] [PMID: 15310732]

[32]   Barreira JCM, Alves RC, Casal S, Ferreira ICFR, Oliveira MBPP, Pereira JA. Vitamin E profile as a reliable authenticity discrimination factor between chestnut (*Castanea sativa* Mill.) cultivars. J Agric Food Chem 2009; 57(12): 5524-8.
       [http://dx.doi.org/10.1021/jf900435y] [PMID: 19489539]

[33]   Barros AIRNA, Nunes FM, Gonçalves B, Bennett RN, Silva AP. Effect of cooking on total vitamin C contents and antioxidant activity of sweet chestnuts (*Castanea sativa* Mill.). Food Chem 2011; 128(1): 165-72.
       [http://dx.doi.org/10.1016/j.foodchem.2011.03.013] [PMID: 25214344]

[34]   Ribeiro B, Rangel J, Valentão P, *et al.* Organic acids in two Portuguese chestnut (*Castanea sativa* Miller) varieties. Food Chem 2007; 100: 504-8.
       [http://dx.doi.org/10.1016/j.foodchem.2005.09.073]

[35]    Venkatraman JT, Chu WC. Effects of dietary ω-3 and ω-6 lipids and vitamin E on serum cytokines, lipid mediators and anti-DNA antibodies in a mouse model for rheumatoid arthritis. J Am Coll Nutr 1999; 18(6): 602-13.
[http://dx.doi.org/10.1080/07315724.1999.10718895] [PMID: 10613412]

[36]    Lee C-YJ, Wan JM-F. Vitamin E supplementation improves cell-mediated immunity and oxidative stress of Asian men and women. J Nutr 2000; 130(12): 2932-7.
[PMID: 11110849]

[37]    Allard JP, Aghdassi E, Chau J, *et al*. Effects of vitamin E and C supplementation on oxidative stress and viral load in HIV-infected subjects. AIDS 1998; 12(13): 1653-9.
[http://dx.doi.org/10.1097/00002030-199813000-00013] [PMID: 9764785]

[38]    Itoh N, Masuo Y, Yoshida Y, Cynshi O, Jishage K, Niki E. γ-Tocopherol attenuates MPTP-induced dopamine loss more efficiently than α-tocopherol in mouse brain. Neurosci Lett 2006; 403(1-2): 136-40.
[http://dx.doi.org/10.1016/j.neulet.2006.04.028] [PMID: 16716512]

[39]    Iqbal K, Khan A, Khattak MMAK. Biological significance of ascorbic acid (vitamin C) in human health – A Review. Pakistan J Nut 2004; 3: 5-13.
[http://dx.doi.org/10.3923/pjn.2004.5.13]

[40]    Rumsey SC, Levine M. Absorption, transport, and disposition of ascorbic acid in humans. J Nutr Biochem 1998; 9: 116-30.
[http://dx.doi.org/10.1016/S0955-2863(98)00002-3]

[41]    Retsky KL, Freeman MW, Frei B. Ascorbic acid oxidation product(s) protect human low density lipoprotein against atherogenic modification. Anti- rather than prooxidant activity of vitamin C in the presence of transition metal ions. J Biol Chem 1993; 268(2): 1304-9.
[PMID: 8419332]

[42]    Carocho M, Barros L, Antonio AL, *et al*. Analysis of organic acids in electron beam irradiated chestnuts (*Castanea sativa* Mill.): Effects of radiation dose and storage time. Food Chem Toxicol 2013; 55: 348-52.
[http://dx.doi.org/10.1016/j.fct.2013.01.031] [PMID: 23376134]

[43]    Silva BM, Andrade PB, Valentão P, Ferreres F, Seabra RM, Ferreira MA. Quince (*Cydonia oblonga* Miller) fruit (pulp, peel, and seed) and Jam: antioxidant activity. J Agric Food Chem 2004; 52(15): 4705-12.
[http://dx.doi.org/10.1021/jf040057v] [PMID: 15264903]

[44]    Vaughan JG, Geissler CA. The new Oxford book of food plants. New York: Oxford University Press 2009.

[45]    Uttara B, Singh AV, Zamboni P, Mahajan RT. Oxidative stress and neurodegenerative diseases: a review of upstream and downstream antioxidant therapeutic options. Curr Neuropharmacol 2009; 7(1): 65-74.
[http://dx.doi.org/10.2174/157015909787602823] [PMID: 19721819]

[46]    Almeida IF, Valentão P, Andrade PB, *et al*. *In vivo* skin irritation potential of a *Castanea sativa* (Chestnut) leaf extract, a putative natural antioxidant for topical application. Basic Clin Pharmacol

Toxicol 2008; 103(5): 461-7.
[http://dx.doi.org/10.1111/j.1742-7843.2008.00301.x] [PMID: 18793273]

[47]    Barreira JCM, Ferreira ICFR, Oliveira MBPP, Pereira JA. Antioxidant activities of the extracts from chestnut flower, leaf, skins and fruit. Food Chem 2008; 107: 1106-13.
[http://dx.doi.org/10.1016/j.foodchem.2007.09.030]

[48]    Živković J, Zeković Z, Mujić I, Tumbas V, Cvetković D, Spasojević I. Antioxidant properties of phenolics in *Castanea sativa* Mill. extracts. Food Technol Biotechnol 2009; 47: 421-7.

[49]    Sapkota K, Park S-E, Kim J-E, *et al.* Antioxidant and antimelanogenic properties of chestnut flower extract. Biosci Biotechnol Biochem 2010; 74(8): 1527-33.
[http://dx.doi.org/10.1271/bbb.100058] [PMID: 20699587]

[50]    Vázquez G, Fernández-Agulló A, Gómez-Castro C, Freire MS, Antorrena G, González-Álvarez J. Response surface optimization of antioxidants extraction from chestnut (*Castanea sativa*) bur. Ind Crops Prod 2012; 35: 126-34.
[http://dx.doi.org/10.1016/j.indcrop.2011.06.022]

[51]    Vázquez G, Fontenla E, Santos J, Freire MS, González-Álvarez J, Antorrena G. Antioxidant activity and phenolic content of chestnut (*Castanea sativa*) shell and eucalyptus (*Eucalyptus globulus*) bark extracts. Ind Crops Prod 2008; 28: 279-85.
[http://dx.doi.org/10.1016/j.indcrop.2008.03.003]

[52]    Barros AIRNA, Nunes FM, Gonçalves B, Bennett RN, Silva AP. Effect of cooking on total vitamin C contents and antioxidant activity of sweet chestnuts (*Castanea sativa* Mill.). Food Chem 2011; 128(1): 165-72.
[http://dx.doi.org/10.1016/j.foodchem.2011.03.013] [PMID: 25214344]

[53]    Carocho M, Antonio AL, Barros L, *et al.* Comparative effects of gamma and electron beam irradiation on the antioxidant potential of Portuguese chestnuts (*Castanea sativa* Mill.). Food Chem Toxicol 2012; 50(10): 3452-5.
[http://dx.doi.org/10.1016/j.fct.2012.07.041] [PMID: 22847131]

[54]    Dinis L-T, Oliveira MM, Almeida J, Costa R, Gomes-Laranjo J, Peixoto F. Antioxidant activities of chestnut nut of *Castanea sativa* Mill. (cultivar 'Judia') as function of origin ecosystem. Food Chem 2012; 132(1): 1-8.
[http://dx.doi.org/10.1016/j.foodchem.2011.09.096] [PMID: 26434256]

[55]    Otles S, Selek I. Phenolic compounds and antioxidant activities of chestnut (*Castanea sativa* Mill.) fruits. Qual Assur Saf Crop 2012; 4: 199-05.
[http://dx.doi.org/10.1111/j.1757-837X.2012.00180.x]

[56]    Delgado T, Malheiro R, Pereira JA, Ramalhosa E. Hazelnut (*Corylus avellana* L.) kernels as a source of antioxidants and their potential in relation to other nuts. Ind Crops Prod 2010; 32: 621-6.
[http://dx.doi.org/10.1016/j.indcrop.2010.07.019]

[57]    Barreira JCM, Ferreira ICFR, Oliveira MBPP, Pereira JA. Antioxidant activity and bioactive compounds of ten Portuguese regional and commercial almond cultivars. Food Chem Toxicol 2008; 46(6): 2230-5.
[http://dx.doi.org/10.1016/j.fct.2008.02.024] [PMID: 18400354]

[58]   Vekiari SA, Gordon MH, García-Macías P, Labrinea H. Extraction and determination of ellagic acid contentin chestnut bark and fruit. Food Chem 2008; 110(4): 1007-11.
[http://dx.doi.org/10.1016/j.foodchem.2008.02.005] [PMID: 26047294]

[59]   De Vasconcelos MCBM, Bennett RN, Quideau S, Jacquet R, Rosa EAS, Ferreira-Cardoso JV. Evaluating the potential of chestnut (*Castanea sativa* Mill.) fruit pericarp and integument as a source of tocopherols, pigments and polyphenols. Ind Crops Prod 2010; 31: 301-11.
[http://dx.doi.org/10.1016/j.indcrop.2009.11.008]

[60]   Veluri R, Singh RP, Liu Z, Thompson JA, Agarwal R, Agarwal C. Fractionation of grape seed extract and identification of gallic acid as one of the major active constituents causing growth inhibition and apoptotic death of DU145 human prostate carcinoma cells. Carcinogenesis 2006; 27(7): 1445-53.
[http://dx.doi.org/10.1093/carcin/bgi347] [PMID: 16474170]

[61]   Hooper L, Kroon PA, Rimm EB, *et al.* Flavonoids, flavonoid-rich foods, and cardiovascular risk: a meta-analysis of randomized controlled trials. Am J Clin Nutr 2008; 88(1): 38-50.
[PMID: 18614722]

[62]   Carocho M, Barreira JCM, Antonio AL, Bento A, Kaluska I, Ferreira ICFR. Effects of electron-beam radiation on nutritional parameters of Portuguese chestnuts (*Castanea sativa* Mill.). J Agric Food Chem 2012; 60(31): 7754-60.
[http://dx.doi.org/10.1021/jf302230t] [PMID: 22809396]

[63]   Fernandes Â, Barreira JCM, Antonio AL, Bento A, Botelho ML, Ferreira ICFR. Assessing the effects of gamma irradiation and storage time in energetic value and in major individual nutrients of chestnuts. Food Chem Toxicol 2011; 49(9): 2429-32.
[http://dx.doi.org/10.1016/j.fct.2011.06.062] [PMID: 21740949]

[64]   Chenlo F, Moreira R, Chaguri L, Torres MD. Effects of storage conditions on sugars and moisture content of whole chestnut fruits. J Food Process Preserv 2010; 34: 609-20.

[65]   Antonio AL, Fernandes Â, Barreira JCM, Bento A, Botelho ML, Ferreira ICFR. Influence of gamma irradiation in the antioxidant potential of chestnuts (*Castanea sativa* Mill.) fruits and skins. Food Chem Toxicol 2011; 49(9): 1918-23.
[http://dx.doi.org/10.1016/j.fct.2011.02.016] [PMID: 21371520]

[66]   Moreira R, Chenlo F, Chaguri L, Vázquez G. Mathematical modelling of the drying kinetics of chestnut (*Castanea sativa* Mill.) Influence of the natural shells. Food Bioprod Process 2005; 83: 306-14.
[http://dx.doi.org/10.1205/fbp.05035]

[67]   Guiné RPF, Fernandes RMC. Analysis of the drying kinetics of chestnuts. J Food Eng 2006; 76: 460-7.
[http://dx.doi.org/10.1016/j.jfoodeng.2005.04.063]

[68]   Koyuncu T, Serdar U, Tosun I. Drying characteristics and energy requirement for dehydration of chestnuts (*Castanea sativa* Mill.). J Food Eng 2004; 62: 165-8.
[http://dx.doi.org/10.1016/S0260-8774(03)00228-0]

[69]   Correia P, Leitão A, Beirão-da-Costa ML. The effect of drying temperatures on morphological and chemical properties of dried chestnuts flours. J Food Eng 2009; 90: 325-32.

[http://dx.doi.org/10.1016/j.jfoodeng.2008.06.040]

[70]   Correia P, Beirão-da-Costa ML. Effect of drying temperatures on starch-related functional and thermal properties of chestnut flours. Food Bioprod Process 2012; 90: 284-94.
[http://dx.doi.org/10.1016/j.fbp.2011.06.008]

[71]   Moreira R, Chenlo F, Torres MD, Rama B. Influence of the chestnuts drying temperature on the rheological properties of their doughs. Food Bioprod Process 2013; 91: 7-13.
[http://dx.doi.org/10.1016/j.fbp.2012.08.004]

[72]   Moreira R, Chenlo F, Chaguri L, Fernandes C. Water absorption, texture, and color kinetics of air-dried chestnuts during rehydration. J Food Eng 2008; 86: 584-94.
[http://dx.doi.org/10.1016/j.jfoodeng.2007.11.012]

[73]   Moreira R, Chenlo F, Chaguri L, Mayor L. Analysis of chestnut cellular tissue during osmotic dehydration, air drying and rehydration Processes. Dry Technol 2011; 29: 10-8.
[http://dx.doi.org/10.1080/07373937.2010.482709]

[74]   Chenlo F, Moreira R, Fernández-Herrero C, Vázquez G. Mass transfer during osmotic dehydration of chestnut using sodium chloride solutions. J Food Eng 2006; 73: 164-73.
[http://dx.doi.org/10.1016/j.jfoodeng.2005.01.017]

[75]   Chenlo F, Moreira R, Fernández-Herrero C, Vázquez G. Experimental results and modeling of the osmotic dehydration kinetics of chestnut with glucose solutions. J Food Eng 2006; 74: 324-34.
[http://dx.doi.org/10.1016/j.jfoodeng.2005.03.002]

[76]   Chenlo F, Moreira R, Fernández-Herrero C, Vázquez G. Osmotic dehydration of chestnut with sucrose: Mass transfer processes and global kinetics modelling. J Food Eng 2007; 78: 765-74.
[http://dx.doi.org/10.1016/j.jfoodeng.2005.11.017]

[77]   Moreira R, Chenlo F, Torres MD, Vázquez G. Effect of stirring in the osmotic dehydration of chestnut using glycerol solutions. LWT 2007; 40: 1507-14.
[http://dx.doi.org/10.1016/j.lwt.2006.11.006]

[78]   Moreira R, Chenlo F, Chaguri L, Oliveira H. Drying of chestnuts (*Castanea sativa* Mill.) after osmotic dehydration with sucrose and glucose solutions. Dry Technol 2007; 25: 1837-45.
[http://dx.doi.org/10.1080/07373930701677686]

[79]   Moreira R, Chenlo F, Chaguri L, Vázquez G. Air drying and colour characteristics of chestnuts pre-submitted to osmotic dehydration with sodium chloride. Food Bioprod Process 2011; 89: 109-15.
[http://dx.doi.org/10.1016/j.fbp.2010.03.013]

[80]   European Union. Official Journal of the European Union. Commission Decision of 18 September 2008 2008. Sep (2008/753/EC)

[81]   UNEP. Montreal Protocol on substances that deplete the ozone layer: Report of the Methyl Bromide Technical Options Committee. Nairobi, Kenya 2006; pp. 1-414. Available from: http://ozone.unep.org/ Assessment_Panels/TEAP/Reports/MBTOC/index.shtml

[82]   Fernandes Â, Antonio AL, Barros L, *et al.* Low dose γ-irradiation as a suitable solution for chestnut (*Castanea sativa* Miller) conservation: effects on sugars, fatty acids, and tocopherols. J Agric Food Chem 2011; 59(18): 10028-33.
[http://dx.doi.org/10.1021/jf201706y] [PMID: 21823582]

[83]    Künsch U, Schärer H, Patrian B, *et al.* Effects of roasting on chemical composition and quality of different chestnut (*Castanea sativa* Mill) varieties. J Sci Food Agric 2001; 81: 1106-12.
[http://dx.doi.org/10.1002/jsfa.916]

[84]    Silva AP, Santos-Ribeiro R, Borges O, Magalhães B, Silva ME, Gonçalves B. Effects of roasting and boiling on the physical and chemical properties of 11 Portuguese Chestnut cultivars (*Castanea sativa* Mill.). CyTA-J Food 2011; 9: 214-9.
[http://dx.doi.org/10.1080/19476337.2010.518249]

# CHAPTER 7

# Bioactive Compounds of Hazelnuts as Health Promoters

Joana S. Amaral [1,2,*], M. Beatriz P.P. Oliveira[1]

[1] *REQUIMTE-LAQV, Departamento de Ciências Químicas, Faculdade de Farmácia, Universidade do Porto, Porto, Portugal*

[2] *ESTiG, Instituto Politécnico de Bragança, Portugal*

**Abstract:** Hazelnut (*Corylus avellana* L.) belongs to the Betulaceae family and is one of the most popular and commonly consumed tree nuts worldwide. Hazelnuts are highly nutritious, providing macronutrients (fat, protein and carbohydrates), micronutrients (vitamins and minerals) and several bioactive phytochemicals. So far, different phytochemicals have been described in hazelnuts, such as phenolic acids, flavonoids, condensed tannins and phytosterols. Among the array of phytochemicals present in hazelnuts, several have been associated with interesting properties such as antioxidant, anti-inflammatory, anti-proliferative and hipocholesterolemic activities thus potentially contributing for beneficial health effects related to hazelnut consumption. Even though hazelnuts have a high fat content, its inclusion has been recommended as part of a healthy-diet. The health benefits of hazelnut consumption have been mainly associated with its favourable lipidic composition and fat-soluble bioactives but also to its content in other compounds such as L-arginine and antioxidant phytochemicals. This chapter aims at providing detailed and up-to-date information on hazelnut bioactive compounds composition and related health aspects, including data from epidemiological and clinical studies.

**Keywords:** Chemical composition, Hazelnut, Health benefits.

* **Corresponding author Joana S. Amaral:** ESTiG, Instituto Politécnico de Bragança, Campus de Sta. Apolònia, 5301-857 Bragança, Portugal; Tel: (+351) 273 303 138; Fax: (+351) 273 325 40; Email: jamaral@ipb.pt.

**Luís Rodrigues da Silva and Branca Maria Silva (Eds.)**

# INTRODUCTION

Hazelnut (*Corylus avellana* L.) is a highly appreciated nut, being mainly cultivated in the Black Sea region of Turkey, in southern European countries, such as Italy, Spain, France, Greece and Portugal, and in the United States of America (USA). Other countries with relevant production of hazelnut include Azerbaijan, Georgia, Iran and China [1]. As happens with most crops, the world annual production of hazelnuts shows some fluctuation depending on the climatic conditions variations from year to year.

In the last decade, overall production of in-shell hazelnut varied among 742,125 tonnes in 2011 and 1069,175 tonnes in 2008 [1]. Turkey is the largest producer of hazelnut, comprising 64% of the world's total production in 2013, followed by Italy with a production corresponding to 11% [1]. Turkish hazelnut is generally divided into two main groups according to the quality of the nut, namely "Giresun" (or prime quality hazelnut) and "Levant" (or secondary quality) [2]. Among the prime quality hazelnuts, which are mainly grown in the Giresun province and neighbouring cities, Tombul cultivar is the most known and requested worldwide due to its high oil content and distinctive organoleptic properties [2, 3]. In Italy, the second world largest producer, hazelnut is mainly grown in the provinces of Campania, Lazio, Piemonte and Sicilia. Currently, "Nocciola Romana", corresponding to the Italian cultivars Tonda Gentile Romana and Nocchione in at least 90% and Tonda di Giffoni and Barretona in less than 10%, is commercialized under PDO (Protected Designation of Origin) designation. Among the Italian hazelnuts, two cultivars have also been attributed with PGI (Protected Geographical Indication) designation, namely "Nocciola di Giffoni", corresponding to cultivar Tonda di Giffoni, and "Nocciola del Piemonte", corresponding to cultivar Tonda Gentile delle Langhe [4]. Additionally, the Spanish "Avellana de Reus", comprising the cultivars Negreta, Pauetet, Gironella, Morella and Culplana, and the Frech "Noisette de Cervione" corresponding to the cultivar Fertille de Coutard, have also been designated with PDO and PGI designations, respectively [4].

The great majority of hazelnuts (~90%) is used in the food industry, as ingredient in chocolates, pastry and bakery, desserts and to add flavour and texture in other formulations, with approximately 10% of world's production being sold as fresh unshelled nuts, to be consumed raw or roasted [5, 6]. Besides the high quality and absence of defects, frequently other requisites mainly related to kernel's physical features are required by the food industry, such as uniformity, size and shape [4]. Due to its characteristics, the cultivars more frequently used in the food industry include the Italian Tonda Gentile delle Langhe, Tonda Gentile Romana, Tonda di Giffoni, San Giovanni and Mortarella, the Turkish Tombul, Sivri, Palaz and Fosa and the Spanish Negret and Pauetet [4]. By the contrary, hazelnuts with large kernels such as cultivars Butler, Ennis and Lansing, which are frequently produced in the USA, are generally considered as being "table hazelnuts" since they are mainly intended to be consumed raw [7].

Hazelnut is considered to be a highly nutritious nut since it is rich in monounsaturated fatty acids (MUFA) and also contains a wide variety of vitamins, minerals and phytochemicals. Besides being regarded as a good source of nutrients, the consumption of hazelnuts has been recently associated with different health benefits. For these reasons, this chapter intends to provide a comprehensive overview on the existing knowledge on hazelnut composition, focusing in what concerns health promoting substances. Where possible, the composition of hazelnut by-products will be also mentioned since they potentially can be used as a source of phytochemicals.

## CHEMICAL COMPOSITION

### Proximate Composition

Up to now, several researchers have examined the proximate composition of several hazelnut cultivars grown in different countries, including Iran, Italy, New Zealand, Portugal, Spain, Turkey and USA [2, 8 - 14]. Although several factors can influence the final composition of hazelnut, including genetics, edaphoclimatic factors, geographical origin and harvesting time, among others, fat is the predominant component of hazelnut, followed by carbohydrate and protein, which sometimes present very close values among each other. Considering the reported values for several cultivars grown in different countries, fat is reported to vary among 43.2-69.0 g/100 g, carbohydrates among 5.8-26.0 g/100 g, protein among 9.3-24.6 g/100 g and ash among 2.1-4.1 g/100 g [2, 8 - 15]. Similarly to other nuts, hazelnut also presents a low moisture content (ranging from 3.5-8.8 g/100 g) which is associated to a low probability of microorganisms' growth and a high shelf life. The influence of geographical origin and edaphoclimatic factors is evidenced when comparing the values for proximate composition of nuts from the same cultivars but growing on distinct location and/or years. As example, cultivars Segorbe and Daviana grown either in 2001 in Portugal or in 2010 in Iran presented very distinct composition, namely Segorbe presented higher values (g/100 g, dry weight) for the nuts grown in Iran in what respects fat (70.5 *vs*. 66.2) and protein (19.4 *vs* 13.0) but lower values for carbohydrates (6.3 *vs* 18.0) compared to those in Portugal, while cultivar Daviana presented much lower fat values for the nuts grown in Iran (46.8 *vs* 64.4) and higher contents of protein (21.4 *vs* 12.0) and carbohydrates (28.2 *vs* 20.6) compared to the same cultivar grown in Portugal [8, 11]. The use of fertilization can also affect proximate composition of nuts. Recently, significant differences in terms of protein, oil and ash contents due to the application of different levels of zinc fertilization were reported for Tombul hazelnut grown in Turkish orchards [16].

## Amino Acids

Hazelnut is generally reported to contain all essential amino acids, although lysine and tryptophan are considered as being limiting amino acids [2, 17]. Nevertheless, some studies referred the absence of tryptophan in Nebraska hybrid hazelnuts [17]. The most abundant amino acids in hazelnut are glutamic acid, arginine and aspartic acid. The non-essential amino acids are in majority corresponding to approximately 68-73% of total amino acids content [2, 9, 19, 20].

The intake of plant protein compared to animal protein has generally been associated with a lower risk of cardiovascular disease (CVD). Although the explanation for this association is still not completely clear, it has been suggested that in part it can be due to the high contents in arginine and low ratio lysine:arginine (Lys:Arg) in several plant foods [21]. Hazelnuts have a high content of arginine, a precursor of nitric oxide which has many activities in the human body such as vasodilation, antioxidative and antiplatelet effects [21, 22]. The administration of arginine in animal models and in humans with hypercholesterolemia and atherosclerosis has resulted in improvement of the endothelial function [23]. Consequently, foods rich in arginine, such as hazelnut, have been associated with cardioprotective benefits. Based on data from participants in the Third National Health and Nutrition Examination Survey (NHANES III) (n = 13401 individuals), Wells *et al.* evaluated the relation between the consumption levels of dietary arginine with the risk of CVD as determined by the levels of C-reactive protein [24]. The obtained results suggested that individuals may be able to decrease their CVD risk by following a diet high in arginine-rich foods, such as nuts and fish [24]. Dietary arginine has also been referred to have the capacity of altering the balance of energy intake and expenditure in favour of fat loss or reduced growth of white adipose tissue by stimulating mitochondrial biogenesis signalling and brown adipose tissue development [25, 26].

Additionally, hazelnuts present a low level of lysine. Considering that this amino acid shares the same *in vivo* uptake transport as arginine, some authors consider that a low ratio Lys:Arg can be associated with a lower risk of atherosclerosis [21]. Nevertheless, in a randomized study comprising 31 hypercholesteraemic individuals, diets differing in Lys:Arg ratio only showed small effects, or none at all, on CVD risk factors and vascular reactivity [27]. More recently, as part of the Healthy Lifestyle in Europe by Nutrition in Adolescence (HELENA) study, Bel-Serrat *et al.* [25] evaluated the relationship between amino acid intake and serum lipid profile in European adolescents, also aiming to assess whether this association was independent of total fat intake. Arginine intake was found to be inversely associated with serum triglycerides (TG) concentration in both boys and girls and with serum total cholesterol (TC), LDL-cholesterol and Apo B/Apo A1 ratio only in girls. However, the authors concluded that those associations were no longer significant when total fat intake was considered as a confounding factor [25].

## Water-Soluble Vitamins

Hazelnut contains different water-soluble vitamins in its composition, including ascorbic acid, niacin, biotin, folic acid, pantothenic acid, vitamins B1, B2 and B6 [2, 28, 29] whose ingestion is important for the normal function of human organism. Among tree nuts, hazelnut is one of the richest in folic acid, which plays an essential role in nucleotide and protein synthesis, among other functions in the human body [18]. Folate deficit, especially in pregnant women, can lead to severe health injuries, including malformations in the fetus, and is associated to increased levels of homocysteine and higher risk of CVD [18]. Gunes *et al.* [29] evaluated the composition in water-soluble vitamins of 12 hazelnut cultivars grown in Turkey and reported that folic acid content ranged from 11 µg/100 g to 91 µg/100.

## Minerals

Different macroelements (*e.g.* potassium, calcium, phosphorous) and microelements (*e.g.* zinc, copper, selenium) are generally considered to have important roles in growth and cell metabolism. Hazelnut has been reported to serve as a good source of several essential and non-essential minerals. However, its composition can vary widely depending on the variety, geographical origin, climatic conditions, soil composition and selected fertilization and irrigation practices [19, 28, 30, 31]. According to the USDA database [13], potassium is the most abundant mineral in hazelnuts (680 mg/100 g) followed by phosphorous (290 mg/100 g), magnesium (163 mg/100 g), calcium (114 mg/100 g), iron (4.7 mg/100 g) and zinc (2.45 mg/100 g). According to the literature [2, 19, 28, 31 - 36] other minerals have also been described to occur in hazelnuts including manganese, copper, selenium, sodium, chromium, boron, lithium, cobalt, molybdenum, vanadium [2, 19, 28, 33] with the presence of lead, nickel and silver also being reported [2, 36]. Besides being considered a good source of several of the mentioned trace minerals, such as copper, manganese and molybdenum, with regard to nutritional aspects, the high potassium to sodium ratio makes hazelnut interesting for diets with a defined electrolytic balance [2].

Hazelnuts are also considered to be an excellent source of the trace mineral selenium, which is considered to play an antioxidant role in cells, protecting cell membranes by preventing free radical generation, thereby decreasing the risk of cancer and disease of the heart and blood vessels [2]. Selenium content of 60 µg/100 g was reported by Alasalvar *et al.* [2] for Tombul hazelnut grown in Turkey. Considering this data, the authors referred that among tree nuts, only

Brazil nuts contained a higher selenium content than that of Tombul hazelnut. Nevertheless, lower values were also reported by Alasalvar *et al.* [33] for five Turkish cultivars including Tombul hazelnut (5.5-8.1 µg/100 g) while higher contents were reported by Dugo *et al.* [37] for Italian hazelnuts (86.5 µg/100 g) and by Simsek and Aykut [36] for different varieties of Turkish hazelnuts (96-139 µg/100 g). According to Alasalvar *et al.* [33] these differences among selenium values may be due to several factors, such as the sensitivity of the equipment or the extraction methods used, as well as sue to environmental and varietal factors which can lead to significant variations.

## LIPIDIC COMPOUNDS

As referred, hazelnut kernels are rich in lipids, nevertheless they present a desirable lipid profile due to a high content of monounsaturated fatty acids (MUFA) and low content of saturated fatty acids (SFA). Besides triacylglycerols, which are the main lipid class, other lipidic compounds are present including phytosterols, lipossoluble vitamins, phospholipids and sphingolipids.

### Fatty Acids

Hazelnuts have different fatty acids (FA) in their composition as part of triacylglycerols, with at least a total of twenty FA already being reported in the literature [38]. Amaral *et al.* [8] identified fifteen FA in a total of 19 evaluated cultivars grown in Portugal, from which oleic acid (C18:1n9) was the major compound (76.7%-82.8%, mean value 80.4%), followed by linoleic (C18:2n6), palmitic (C16:0), stearic (C18:0) and vaccenic (C18:1n7) acids with mean values of 9.2%, 5.6%, 2.7% and 1.4%, respectively. All other FA were present in amounts lower than 1% [8]. The individual FA content reported for Portuguese samples was, in general terms, in good agreement with those reported in the literature for different hazelnuts cultivars/different geographical origin [10, 12, 15, 20, 31, 39 - 43]. Roasted hazelnuts are reported to suffer minor changes in FA composition, even though the FA profile remains identical to those of the raw nuts [38, 44].

Nowadays, it is generally accepted that, additionally to the amount of total fat, the type of fatty acids that are consumed is also an important issue when health is concerned. Hazelnut presents MUFA as the main FA group while having a low content of SFA. Polyunsaturated FA (PUFA) are frequently in slightly higher contents compared to SFA, but in a similar order of magnitude.

Bacchetta *et al.* [45] analysed the FA composition of 75 European hazelnut cultivars grown in 6 different countries and reported a mean value of 80.9% for MUFA, 10.7% for PUFA and 8.4% for SFA. Compared with other nuts and different vegetable oils, hazelnut presents one of the highest content of MUFA and lowest of SFA [20, 46]. According to the Food and Agriculture Organization (FAO) recommended dietary intakes for total fat and FA intake, dietary fat should provide 20-35% of total energy from which SFA should correspond to a maximum of 10% [47]. SFA, in particular lauric, myristic and palmitic acids have been showed to increase low-density lipoprotein (LDL) cholesterol, therefore increasing the risk of CVD. For this reason, the replacement SFA with PUFA and/or MUFA, whose consumption has been associated with health benefits, has been recommended to decrease the risk of coronary heart disease [47]. Considering the high MUFA content and low SFA content, in particular that of palmitic acid, hazelnut is considered to have a favourable FA profile associated to the health benefits.

**Phytosterols**

Phytosterols and their saturated counterparts, phytostanols, have been showed to have different health benefits among which the inhibition of cholesterol absorption is the most widely known [48]. Phytosterols have also been associated with risk decrease of certain types of cancer [49, 50], decrease risk of CVD [51] and enhance of immune functions [52].

A total of fourteen phytosterols and phytostanols, with amounts ranging from 105.7–195.2 mg/100g of oil, were reported by Crews *et al.* [53] when evaluating authentic hazelnut oils obtained from nuts collected from five countries. The major compounds were β-sitosterol (85.5-161.2 mg/100g of oil), followed by campesterol (5.8-10.7 mg/100g of oil), β-sitostanol (2.9-11.7 mg/100g of oil), Δ5-avenasterol (0.8-9.4 mg/100g of oil), clerosterol (1.5-3.8 mg/100g of oil) and stigmasterol (1.4-5.4 mg/100g of oil). A similar profile was also reported by

Amaral *et al.* [8] when analysing 19 cultivars grown in Portugal. In this case, total amounts ranged from 133.8 to 263.0 mg/100g of oil. The chemical structures of main hazelnut phytosterols and phytostanols are presented on Fig. (1).

Considering the described effects of phytosterols and phytostanols and that the consumption of these compounds in moderate to high doses have been shown to favourably alter whole-body cholesterol metabolism in a dose-dependent manner [54], the health benefits associated with the consumption of hazelnuts may be, at least in part, due to its rich composition in sterols.

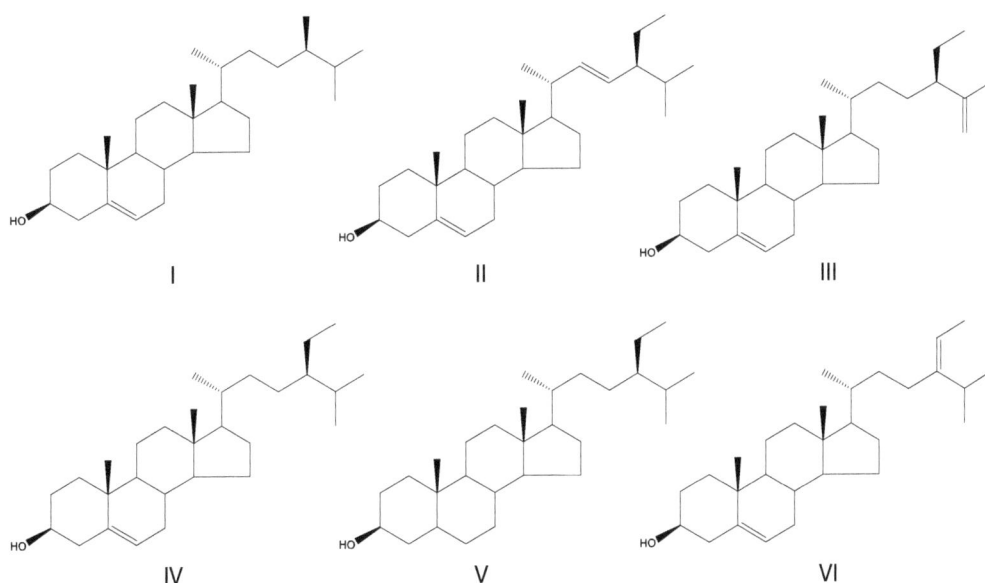

**Fig. (1).** Chemical structures of major phytosterols and phytostanols in hazelnuts (**I:** campesterol; **II:** stigmasterol; **III:** clerosterol; **IV:** *β*-sitosterol; **V:** *β*-sitostanol ; **VI:** $\Delta^5$-avenasterol).

## Vitamin E

Vitamin E encompasses a group of eight naturally occurring fat-soluble compounds, four tocopherols and four tocotrienols (in both cases designated as α, β, γ and δ). Besides being involved in a diversity of physiological and biochemical functions, vitamin E, due to its antioxidant activity, is particularly important in the prevention of lipid oxidation processes. Among the different vitamers, α-tocopherol ($\alpha T$), the predominant form in the living tissues, is the most known and studied one, with several studies demonstrating that it acts as a potent radical scavenging antioxidant, especially in combination with vitamin C. However, recent studies suggest that other vitamers, such as $\gamma$- tocopherols ($\gamma T$) and $\delta$-tocopherols ($\delta T$) and $\gamma$-tocotrienol, have unique antioxidant and anti-inflammatory properties that can even be superior to that of $\alpha T$ in prevention against chronic diseases [55]. Other interesting biological activities have been associated with these compounds, such as neuroprotective, anti-cancer and cholesterol lowering properties, among others [56, 57]. Consequently, since different vitamin E vitamers can play a role in contributing to the total bioactivity in foods, in recent studies dealing with vitamin E composition of several food matrices, all vitamers are analysed instead of just focusing on $\alpha T$.

Hazelnut oil is considered as being an excellent source of vitamin E, with seven tocopherols and tocotrienols already been reported to occur in hazelnuts [58, 59]. The main compound in hazelnut kernels and oil is $\alpha T$ although its content can vary according to different factors such as cultivar and geographical origin. Amaral *et al.* [58] reported a mean value of $\alpha T$ of 29.9 mg/100 g of extracted oil when evaluating 19 cultivars grown in Portugal (ranging from 15.9-38.8 mg/100 g oil), while higher values have been reported for Turkish hazelnuts (ranging from 36.3-46.9 mg/100 g oil) [33]. Among Turkish cultivars, Tombul hazelnut is reported has having the highest $\alpha T$ content [33]. Alasalvar *et al.* [59] reported that among the identified compounds in Tombul cultivar, $\alpha T$ was the most abundant (40.40 mg/100 g) accounting for almost 80% of total vitamin E, followed by $\gamma T$ (8.33 mg/100 g), $\beta$-tocopherol (1.53 mg/100 g) and a small amount of $\delta T$ (0.53 mg/100 g). Tocotrienols contributed only 1.02% to the total vitamin E.

Comparatively with other nuts, hazelnut has been reported has having the highest vitamin E content [60, 61]. In general, the consumption of hazelnuts in the amount recommended by the Food and Drug Administration for nuts [62] can provide approximately 80% of the recommended dietary allowance (RDA) of vitamin E per day [61].

## PHENOLIC COMPOUNDS

Phenolic compounds, including phenolic acids, flavonoids and tannins, are phytochemicals that can play important health benefits due to their antioxidant activity and other interesting properties.

Nowadays, it is of general agreement that the excess of reactive oxygen species (ROS) can lead to oxidative stress that can damage lipids, proteins and nucleic acids in the cells. Therefore, free radical mediated oxidative stress has been associated with the increased risk of several diseases such as CVD, diabetes and certain cancers [18]. Natural antioxidants, such as polyphenols, have the ability of scavenging free radicals and additionally can function as reducing agents, chelator of pro-oxidant metals or as quenchers of singlet oxygen [63], thus can counterbalance and reduce oxidative damage of cellular biomolecules. In this sense, these compounds have been associated to the prevention and/or reduction of different diseases.

### Phenolic Acids

Phenolic acids are naturally present in higher plants, generally as free compounds or as glycosides, esters or in insoluble-bound forms in complex mixtures. Phenolic acids are considered to be good antioxidants, although its activity can vary according to the chemical structure, in particular with the number of hydroxyl and methoxy groups existing in the molecule, with antioxidant activity *in vitro* improving as the number of such substituents increase [64]. Andreasen *et al.* [65] reported that human LDL-cholesterol was better protected against oxidation by caffeic acid, followed by sinapic, ferulic and *p*-coumaric acids.

Different phenolic acids have been reported in hazelnut kernels and also in hazelnut by-products such as tree leaf, green leafy cover and shell. The content of phenolic compounds in hazelnut kernels can vary depending if the nut is evaluated with or without skin since the hazelnut skin is considered as being an interesting source of phenolic compounds [67 - 69]. During hazelnut roasting the skin is generally separated from the kernel and normally discarded, however incomplete removal can occur. Hazelnut skin removal decreases total phenolic content and antioxidant potential as stated by Schimitzer *et al.* [70]. Additionally, the cultivar and the solvents used to prepare the extracts have been reported to exert a great influence on the variability and concentration of phenolic acids [66] as well as on total phenol content and antioxidant activity [67]. Different free and/ or bound phenolic acids have been reported in hazelnut kernels including benzoic and cinnamic acids derivatives, namely caffeic, *p*-coumaric, ferulic, sinapic, gallic, protocatechuic, vanillic, quinic and syringic acids [68, 71 - 74]. Vanillic acid was described has being the major free phenolic acid in aqueous methanolic extracts prepared from deffated dry hazelnut kernels without skin from 15 Turkish cultivars (ranging from 20-255 µg/g) [73] while Shahidi *et al.* [68] reported the presence of *p*-coumaric acid as being the major phenolic acid (free and esterified) in ethanol extract obtained from Tombul hazelnut kernels without skin (208 µg/g of extracts). Ciemniewska-Zytkiewicz *et al.* [40] evaluated phenolic compounds composition in two cultivars grown in Poland (hazelnut kernels with skin) and reported the presence of two free phenolic acids in ethanolic extracts, gallic and protocatechuic acids, and six bounded phenolic acids, namely protocatechuic, vanillic, sinapic, *p*-coumaric and quinic acids, with gallic acid accounting approximately from 30 to 40% of total free phenolic acids and caffeic acid has being the major bounded phenolic acid (accounting from 40 to 72% of total bounded phenolic acids).

As referred, phenolic acids profile has also been investigated in different hazelnut by-products due to the possible interest on its use as potential sources of phytochemicals [40, 68, 74 - 77]. Among by-products, hazelnut shell has been described as representing a rich source of natural phenolic antioxidants [67, 72]. According to Shahidi *et al.* [68] the highest content of phenolic acids in Tombul cultivar was found for hazelnut shell, followed by the green leafy cover, tree leaf and hazelnut skin. All evaluated by-products presented higher phenolic acids content compared to the hazelnut kernel. Gallic acid was the major phenolic acid in hazelnut shell which is in good agreement with the study of Ciemniewska-Zytkiewicz *et al.* [40] that reported gallic acid has major compound among the five detected free phenolic acids in hazelnut shell. The composition of phenolic compounds in hazelnut tree leaf has also been evaluated by Amaral *et al.* [75, 76] who reported the identification and quantification of four phenolic acids, namely 3-caffeoylquinic acid, 5-caffeoylquinic acid, caffeoyltartaric acid, and *p*-

coumaroyltartaric acid, in hazelnut leaves from different cultivars grown in Portugal.

**Flavonoids**

Flavonoids, the most common and widely distributed group of plant phenolics, comprise a large number of compounds included in different classes (flavonols flavanols, flavones, flavanones, isoflavones, anthocyanins, and others) depending on the structure of the compound. Flavonoids are ubiquous secondary plant metabolites that, during the last decades, have been under intensive investigation owing to their putative health properties. In this respect, different epidemiologic studies suggest a protective role of dietary flavonoids against coronary heart disease and CVD [78 - 80].

Flavonoids are mainly known for their recognized antioxidant activity towards different radicals, which is mostly due to the redox properties of their phenolic hydroxyl groups and structural relationships between different parts of their chemical structure [78]. Flavonoids in general have been shown to be highly effective scavengers of most types of oxidising molecules, including singlet oxygen and various free radicals, which are involved in cell damage and possibly related with several diseases such as cancer. Moreover, *in vitro* experimental data also pointed that flavonoids can exert different activities such as antiinflamatory, antiallergic, antiviral and anticarcinogenic properties [78, 81].

Ciarmiello *et al.* [3] evaluated 29 hazelnut cultivars using HPLC-UV and electrospray ionization multistage ion trap mass spectrometry (ESI-ITMS$^n$) and reported the presence flavan-3-ols such as catechin, and compounds thereafter such as procyanidin B2, six procyanidin oligomers (including three unidentified dimers and three unidentified procyanidin trimers), flavonols (quercetin-3-*O*-rhamnoside and myricetin-3-*O*-rhamnoside) and one dihydrochalcone (phloretin-2-*O*-glucoside) in all the analyzed cultivars. Quercetin-3-*O*-rhamnoside was the major flavonol, ranging from 1.9 to 6.6 mg/kg of fresh kernel. Similar results were reported by Jakopic *et al.* [77] who studied the phenolic composition of hazelnut kernels from 20 different cultivars grown in Slovenia using HPLC-MS$^n$ and reported the presence of flavan-3-ols and flavonols among other phenolics, with the major group in hazelnut kernels being the flavan-3-ols ((+)-catechin and (-)-epicatechin). In good agreement with previous studies, more recently, twelve phenolic compounds including catechin, epicatechin, 2 procyanidin dimmers, 2 procyanidin trimers, 3 flavonols and 1 chalcone were identified and quantified by Ciemniewska-Zytkiewicz [40]. Among those, the identified procyanidin dimmer varied between 5.3 and 10.5 µg/g of dry weight kernel and quercetin-3-*O*-rhamnoside between 4.9 and 2.8 µg/g of dry weight kernel in Webba Cenny and Kataloński cultivars, respectively.

## Tannins

Tannins are generally divided in two main groups, hydrolysable tannins (comprising gallotannins and ellagitannins) and condensed tannins (also known as proanthocyanidins). Proanthocyanidins containing (epi)catechin, (epi)afzelechin or (epi)gallocatechin as flavan-3-ol units, are named procyanidins, propelargonidins or prodelphinidins, respectively [82]. Proanthocyanidins are interesting compounds due to their high antioxidant activity and possible effects on human health by reducing the risk of CVD and cancer [83]. It has also been demonstrated that these compounds increase plasma antioxidant activity, have a positive effect on vascular function, reduce platelet activity and protect the intestinal mucosa against oxidative stress [84, 85]. Additionally, different studies, mainly performed with grape seed proanthicyanidins have demonstrated that these compounds can act as anticarcinogenic agents due to their antioxidant, apoptosis-inducing and immuno-modulation activities, among others [86].

As referred in the previous section, together with phenolic acids and flavonols, different procyanidin oligomers have been described in hazelnut kernels. Recently several flavan 3-ols derivatives have been identified and quantified in hazelnut kernels, namely three procyanidin dimers, six procyanidin trimers, three procyanidin tetramers, together with catechin, epicatechin and epicatechin 3-gallate [82]. As reported by Monagas *et al.* [87] hazelnut skin is also rich in proanthocyanidins, mostly of B-type with a degree of polymerization (DP) up to 7 but also being detected procyanidin–prodelphinidin heteropolymers up to DP9. Recently, two new A-type dimeric prodelphinidins, epigallocatechin-(2β→O7, 4β→8)-catechin and epigallocatechin-(2β→O5, 4β→6)-catechin, were isolated from the skins of roasted hazelnut and reported to occur as minor compounds [88]. Hazelnuts have been reported to contain a high content of total proanthocianydins presenting the highest amount when compared to other tree nuts [89].

Also recently, the presence of hydrolyzable tannins in hazelnuts was reported for the first time, comprising the B type dimer gallate, glansreginin A (ranging from 18.0 to 110.9 mg/kg and glansreginin B (ranging from 9.6 to 56.1 mg/kg) [82].

## HEALTH ASPECTS: EPIDEMIOLOGICAL AND CLINICAL STUDIES

So far, different studies demonstrated a significant inverse association between tree nuts consumption and all-cause mortality and death due to heart disease [90]. In fact, in several feeding studies, the consumption of nuts has been associated with reduced levels of different CVD risk factors including total cholesterol, LDL-cholesterol, triglycerides, apolipoprotein A and B, markers of oxidative stress, insulin resistance, *etc.* [90]. However, in some studies inconsistent results were achieved and in others such association was not always evident [90, 91]. According to O'Neil *et al.* [90] the conflicting results in some feeding and epidemiological studies could be due to factors related to the studied population, the type and amount of consumed nuts, the length of the feeding trial, among others. Thus, to clarify this question, the authors evaluated the data obtained from participants of the National Health and Nutrition Examination Survey (NHANES) 2005–2010 (n = 14,386; 51 % males) and concluded that individuals that consumed tree nuts had better weight/adiposity measures and a lower risk of obesity, overweight/obesity, and elevated waist circumference than non-consumers [90].

So far, several clinical studies are available on the literature to investigate the effects of diet supplementation with different nuts. Nevertheless, studies specifically including hazelnuts in the human diet are still scarce [92 - 96]. In the study of Durak *et al.* [92], thirty healthy and young volunteers consumed hazelnuts (1g/day/kg body weight) in addition to their normal daily diets for 30 days. Blood samples collected before and after this period demonstrated that total cholesterol, LDL-cholesterol and malondialdehyde were lower and that high density lipoprotein (HDL) cholesterol, triglyceride levels and antioxidant potential were higher after hazelnut supplementation. Latter, the effect of a hazelnut-enriched diet (40g/day) on plasma cholesterol and lipoprotein profiles in hypercholesterolemic adult men (n=15) was evaluated by comparing with baseline and control diet [93]. The authors concluded that, compared with baseline, hazelnut consumption favourably decreased very-low density lipoproteins (VLDL) cholesterol, triglycerides, apolipoprotein B, while increasing HDL. A decreasing trend was also observed for other parameters such as total and LDL cholesterol, though not being statistically significant. Additionally, in this study, the subject's body weights remained stable throughout the study which is in good agreement with previous data that suggests that adding nuts to the habitual diet does not induce weight gain and may even help losing it [94].

In the work performed by Yücesan *et al.* [95] a single intervention study design (4 week period consuming a with-skin-hazelnut-enriched diet of 1g/kg/day) was used to determine the effects of diet supplementation with hazelnuts on the LDL atherogenic tendency by evaluating several different parameters. The authors concluded that hazelnut can play an important role by lowering the susceptibility of LDL to oxidation and the levels of plasma oxidized-LDL while increasing the ratio of large/small LDL beyond its beneficial effects on lipid and lipoprotein levels [95]. Recently, Orem *et al.* assessed the antiatherogenic effect of hazelnut consumption in hypercholesterolemic individuals by evaluating cardiovascular risk biomarkers, inflammatory markers and blood biochemical parameters [96]. Results showed a decrease on triglycerides, total and LDL-cholesterol and an increase of HLD, apo AI and vitamin E levels in plasma and LDL particles after hazelnut-enriched diet. Vascular endothelial function was also improved upon hazelnut consumption, but this change was reversible when hazelnut consumption was stopped. According to the authors the observed beneficial effects of hazelnut-enriched diet on cardiovascular risk markers and endothelial function may be attributed to hazelnut bioactive compounds consumption, including MUFA, vitamin E, L-arginine and folic acid, among others [96]. The potential for hazelnuts to improve blood lipids was also reported by Tey *et al.* [97] when evaluating the impact of the physical form of the nut (ground, sliced and whole nut) on the hypocholesterolemic effect obtained due to the consumption of a hazelnut-enriched diet. The improvement of plasma lipoprotein and α-tocopherol concentration was similar among mildly hypercholesterolemic subjects with no influence regarding the type of mechanical processing of the nut.

## CONCLUDING REMARKS

Nuts in general are frequently included in healthy diets such as the Mediterranean diet. Hazelnut, in particular, present a favourable composition, being rich in MUFA and low in SFA, presenting with high levels of vitamin E (having α-tocopherol as major compound but also having other vitamers including tocotrienols), and having a high content of phytosterols and of several antioxidant phytochemicals. Hazelnuts also present an interesting composition regarding water-soluble vitamins and minerals, providing iron, zinc, copper, calcium, potassium, magnesium, selenium, folate, among others, while having a high K/Na ratio, desirable for electrolyte balance. So far, different feeding studies have showed that hazelnut consumption may have a positive influence on human health, in particular in lowering CVD risk, which may be attributed to its interesting composition. Therefore, taking into account the up-to-date knowledge, hazelnuts, if possible with skin, should be consumed as part of a healthy diet.

## CONFLICT OF INTEREST

The authors confirm that they have no conflict of interest to declare for this publication.

## ACKNOWLEDGEMENTS

To FCT grant no. LAQV UID/QUI/50006/2013.

## REFERENCES

[1]     FAOSTAT. Available from: http://faostat.fao.org/site/567/DesktopDefault.aspx?PageID=567#ancor acessed June 2015

[2]     Alasalvar C, Shahidi F, Liyanapathirana CM, Ohshima T. Turkish Tombul hazelnut (*Corylus avellana* L.). 1. Compositional characteristics. J Agric Food Chem 2003; 51(13): 3790-6. [http://dx.doi.org/10.1021/jf0212385] [PMID: 12797745]

[3]     Ciarmiello LF, Mazzeo MF, Minasi P, *et al.* Analysis of different European hazelnut (*Corylus avellana* L.) cultivars: authentication, phenotypic features, and phenolic profiles. J Agric Food Chem 2014; 62(26): 6236-46.

[http://dx.doi.org/10.1021/jf5018324] [PMID: 24927513]

[4]    European Commission. Agriculture and Rural Development, Database Of Origin & Registration (DOOR). Available from: http://ec.europa.eu/agriculture/quality/door/list.html 2015. Acessed June 2015

[5]    Ozdemir F, Akinci I. Physical and nutritional properties of four major commercial Turkish hazelnut varieties. J Food Eng 2004; 63(3): 341-7.
       [http://dx.doi.org/10.1016/j.jfoodeng.2003.08.006]

[6]    Valentini N, Rolle L, Stévigny C, Zeppa G. Mechanical behaviour of hazelnuts used for table consumption under compression loading. J Sci Food Agric 2006; 86: 1257-62.
       [http://dx.doi.org/10.1002/jsfa.2486]

[7]    Bergougnoux F, Germain E, Sarraquigne JP, Eds. Le noisetier: production et culture. Influvec. 1978. Available from: http://www.amazon.fr/Le-noisetier-Production-culture-Bergougnoux/dp/B0000DXJ6I

[8]    Amaral JS, Casal S, Citova I, Santos A, Seabra RM, Oliveira BP. Characterization of several hazelnut (*Corylus avellana* L.) cultivars based in chemical, fatty acid and sterol composition. Eur Food Res Technol 2006; 222(3-4): 274-80.
       [http://dx.doi.org/10.1007/s00217-005-0068-0]

[9]    Savage GP, McNeil DL. Chemical composition of hazelnuts (*Corylus avellana* L.) grown in New Zealand. Int J Food Sci Nutr 1998; 49(3): 199-203.
       [http://dx.doi.org/10.3109/09637489809086412] [PMID: 10616661]

[10]   Ruggeri S, Cappelloni M, Gambelli L, Carnovale E. Chemical composition and nutritive value of nuts grown in Italy. Ital J Food Sci 1998; 10(3): 243-52.

[11]   Hosseinpour A, Seifi E, Javadi D, Ramezanpour SS, Molnar TJ. Nut and kernel characteristics of twelve hazelnut cultivars grown in Iran. Sci Hortic (Amsterdam) 2013; 153: 157-7.
       [http://dx.doi.org/10.1016/j.scienta.2013.01.021]

[12]   Parcerisa J, Boatella J, Codony R, *et al.* Comparison of fatty-acid and triacylglycerol compositions of different hazelnut varieties (*Corylus avellana* L.) cultivated in Catalonia (Spain). J Agric Food Chem 1995; 43(1): 13-6.
       [http://dx.doi.org/10.1021/jf00049a004]

[13]   US Department of Agriculture (USDA). USDA National Nutrient Database for Standard Reference Release 27, Basic Report: 12120, Nuts, hazelnuts or filberts. Available from: http://ndb.nal.usda.gov /ndb/foods/show/3698?manu=&fgcd           =           http://ndb.nal.usda.gov/ndb/foods/show /3698?manu=&fgcd=#id-a Acessed June 2015

[14]   Gunes NT, Koksal AI, Artik N, Poyrazoglu E. Biochemical Content of Hazelnut (*Corylus avellana* L.) Cultivars from West Black Sea Region of Turkey. Eur J Hortic Sci 2010; 75(2): 77-84.

[15]   Locatelli M, Coisson JD, Travaglia F, *et al.* Chemotype and genotype chemometrical evaluation applied to authentication and traceability of "Tonda Gentile Trilobata" hazelnuts from Piedmont (Italy). Food Chem 2011; 129(4): 1865-73.
       [http://dx.doi.org/10.1016/j.foodchem.2011.05.134]

[16]   Özenç N, Bender Özenç D. Nut traits and nutritional composition of hazelnut (*Corylus avellana* L.) as influenced by zinc fertilization. J Sci Food Agric 2015; 95(9): 1956-62.

[http://dx.doi.org/10.1002/jsfa.6911] [PMID: 25224327]

[17]    Xu YX, Hanna MA. Evaluation of Nebraska hybrid hazelnuts: Nut/kernel characteristics, kernel proximate composition, and oil and protein properties. Ind Crops Prod 2010; 31(1): 84-91.
[http://dx.doi.org/10.1016/j.indcrop.2009.09.005]

[18]    Sathe SK, Monaghan EK, Kshirsagar HH, Venkatachalam M. Chemical composition of edible nut seeds and its implications in Human health. In: Alasalvar C, Shahidi F, Eds. Tree nuts Composition, phytochemicals and health effects. Boca Raton: CRC Press 2009; pp. 11-36.

[19]    Koksal AH, Artik N, Simsek A, Gunes N. Nutrient composition of hazelnut (*Corylus avellana* L.) varieties cultivated in Turkey. Food Chem 2006; 99(3): 509-15.
[http://dx.doi.org/10.1016/j.foodchem.2005.08.013]

[20]    Venkatachalam M, Sathe SK. Chemical composition of selected edible nut seeds. J Agric Food Chem 2006; 54(13): 4705-14.
[http://dx.doi.org/10.1021/jf0606959] [PMID: 16787018]

[21]    Brufau G, Boatella J, Rafecas M. Nuts: source of energy and macronutrients. Br J Nutr 2006; 96 (Suppl. 2): S24-8.
[http://dx.doi.org/10.1017/BJN20061860] [PMID: 17125529]

[22]    Liu X, Wang M, Zhang N, Fan Z, Fan Y, Deng X. Effects of endothelium, stent design and deployment on the nitric oxide transport in stented artery: a potential role in stent restenosis and thrombosis. Med Biol Eng Comput 2015; 53(5): 427-39.
[http://dx.doi.org/10.1007/s11517-015-1250-6] [PMID: 25715753]

[23]    Gornik HL, Creager MA. Arginine and endothelial and vascular health. J Nutr 2004; 134(10) (Suppl.): 2880S-7S.
[PMID: 15465805]

[24]    Wells BJ, Mainous AG III, Everett CJ. Association between dietary arginine and C-reactive protein. Nutrition 2005; 21(2): 125-30.
[http://dx.doi.org/10.1016/j.nut.2004.03.021] [PMID: 15723738]

[25]    Bel-Serrat S, Mouratidou T, Huybrechts I, *et al.* The role of dietary fat on the association between dietary amino acids and serum lipid profile in European adolescents participating in the HELENA Study. Eur J Clin Nutr 2014; 68(4): 464-73.
[http://dx.doi.org/10.1038/ejcn.2013.284] [PMID: 24495993]

[26]    McKnight JR, Satterfield MC, Jobgen WS, *et al.* Beneficial effects of L-arginine on reducing obesity: potential mechanisms and important implications for human health. Amino Acids 2010; 39(2): 349-57.
[http://dx.doi.org/10.1007/s00726-010-0598-z] [PMID: 20437186]

[27]    Vega-López S, Matthan NR, Ausman LM, *et al.* Altering dietary lysine:arginine ratio has little effect on cardiovascular risk factors and vascular reactivity in moderately hypercholesterolemic adults. Atherosclerosis 2010; 210(2): 555-62.
[http://dx.doi.org/10.1016/j.atherosclerosis.2009.12.002] [PMID: 20042191]

[28]    Ackurt F, Ozdemir M, Biringen G, Loker M. Effects of geographical origin and variety on vitamin and mineral composition of hazelnut (*Corylus avellana* L.) varieties cultivated in Turkey. Food Chem 1999; 65(3): 309-13.

[http://dx.doi.org/10.1016/S0308-8146(98)00201-5]

[29]  Gunes NT, Koksal AI, Artik N, Poyrazoglu E. Biochemical content of hazelnut (*Corylus avellana* L.) cultivars from West Black Sea Region of Turkey. Eur J Hortic Sci 2010; 75(2): 77-84.

[30]  Ozenc N, Ozenc DB. Effect of iron fertilization on nut traits and nutrient composition of 'Tombul' hazelnut (*Corylus avellana* L.) and its potential value for human nutrition. Acta Agr Scand B-S P 2014; 64(7): 633-43.

[31]  Ozdemir M, Ackurt F, Kaplan M, *et al.* Evaluation of new Turkish hybrid hazelnut (*Corylus avellana* L.) varieties: fatty acid composition, alpha-tocopherol content, mineral composition and stability. Food Chem 2001; 73(4): 411-5.
[http://dx.doi.org/10.1016/S0308-8146(00)00315-0]

[32]  Parcerisa J, Rafecas M, Castellote AI, *et al.* Influence of variety and geographical origin on the lipid fraction of hazelnuts (*Corylus avellana* L.) from Spain. 3. Oil Stability, tocopherol content and some mineral contents (Mn, Fe, Cu). Food Chem 1995; 53(1): 71-4.
[http://dx.doi.org/10.1016/0308-8146(95)95789-9]

[33]  Alasalvar C, Amaral JS, Satir G, Shahidi F. Lipid characteristics and essential minerals of native Turkish hazelnut varieties (*Corylus avellana* L.). Food Chem 2009; 113(4): 919-25.
[http://dx.doi.org/10.1016/j.foodchem.2008.08.019]

[34]  Xu YX, Hanna MA. Nutritional and anti-nutritional compositions of defatted Nebraska hybrid hazelnut meal. Int J Food Sci Technol 2011; 46(10): 2022-9.
[http://dx.doi.org/10.1111/j.1365-2621.2011.02712.x]

[35]  Vujevic P, Petrovic M, Vahcic N, Milinovic B, Cmelik Z. lipids and minerals of the most represented hazelnut varieties cultivated in Croatia. Ital J Food Sci 2014; 26(1): 24-30.

[36]  Simsek A, Aykut O. Evaluation of the microelement profile of Turkish hazelnut (*Corylus avellana* L.) varieties for human nutrition and health. Int J Food Sci Nutr 2007; 58(8): 677-88.
[http://dx.doi.org/10.1080/09637480701403202] [PMID: 17852487]

[37]  Dugo G, La Pera L, Lo Turco V, Mavrogeni E, Alfa M. Determination of selenium in nuts by cathodic stripping potentiometry (CSP). J Agric Food Chem 2003; 51(13): 3722-5.
[http://dx.doi.org/10.1021/jf021256m] [PMID: 12797733]

[38]  Alasalvar C, Pelvan E, Topal B. Effects of roasting on oil and fatty acid composition of Turkish hazelnut varieties (*Corylus avellana* L.). Int J Food Sci Nutr 2010; 61(6): 630-42.
[http://dx.doi.org/10.3109/09637481003691820] [PMID: 20384549]

[39]  Alasalvar C, Shahidi F, Ohshima T, *et al.* Turkish Tombul hazelnut (*Corylus avellana* L.). 2. Lipid characteristics and oxidative stability. J Agric Food Chem 2003; 51(13): 3797-805.
[http://dx.doi.org/10.1021/jf021239x] [PMID: 12797746]

[40]  Ciemniewska-Żytkiewicz H, Verardo V, Pasini F, Bryś J, Koczoń P, Caboni MF. Determination of lipid and phenolic fraction in two hazelnut (*Corylus avellana* L.) cultivars grown in Poland. Food Chem 2015; 168: 615-22.
[http://dx.doi.org/10.1016/j.foodchem.2014.07.107] [PMID: 25172755]

[41]  Kiralan S, Yorulmaz A, Simsek A, Tekin A. Classification of Turkish hazelnut oils based on their triacylglycerol structures by chemometric analysis. Eur Food Res Technol 2015; 240(4): 679-88.

[http://dx.doi.org/10.1007/s00217-014-2371-0]

[42]    Savage GP, McNeil DL, Dutta PC. Lipid composition and oxidative stability of oils in hazelnuts (*Corylus avellana* L) grown in New Zealand. J Am Oil Chem Soc 1997; 74(6): 755-9.
[http://dx.doi.org/10.1007/s11746-997-0214-x]

[43]    Parcerisa J, Richardson DG, Rafecas M, Codony R, Boatella J. Fatty acid distribution in polar and nonpolar lipid classes of hazelnut oil (*Corylus avellana* L.). J Agr. Food Chem 1997; 45(10): 3887-90.
[http://dx.doi.org/10.1021/jf9703112]

[44]    Amaral JS, Casal S, Seabra RM, Oliveira BP. Effects of roasting on hazelnut lipids. J Agric Food Chem 2006; 54(4): 1315-21.
[http://dx.doi.org/10.1021/jf052287v] [PMID: 16478254]

[45]    Bacchetta L, Aramini M, Zini A, *et al.* Fatty acids and alpha-tocopherol composition in hazelnut (*Corylus avellana* L.): a chemometric approach to emphasize the quality of European germplasm. Euphytica 2013; 191(1): 57-73.
[http://dx.doi.org/10.1007/s10681-013-0861-y]

[46]    Codex Alimentarius. Codex Standard for named vegetable oils. CODEX STAN (210) 1999. Available from: http://www.fao.org/docrep/004/y2774e/y2774e04.htm#bm4.1

[47]    FAO. Fats and fatty acids in human nutrition. Report of an expert consultation FAO Food and Nutrition Paper 91, 2010; 1-19.

[48]    Kritchevsky D, Chen SC. Phytosterols - health benefits and potential concerns: a review. Nutr Res 2005; 25(5): 413-28.
[http://dx.doi.org/10.1016/j.nutres.2005.02.003]

[49]    Woyengo TA, Ramprasath VR, Jones PJ. Anticancer effects of phytosterols. Eur J Clin Nutr 2009; 63(7): 813-20.
[http://dx.doi.org/10.1038/ejcn.2009.29] [PMID: 19491917]

[50]    Jones PJ, AbuMweis SS. Phytosterols as functional food ingredients: linkages to cardiovascular disease and cancer. Curr Opin Clin Nutr Metab Care 2009; 12(2): 147-51.
[http://dx.doi.org/10.1097/MCO.0b013e328326770f] [PMID: 19209468]

[51]    Hooper L, Summerbell CD, Higgins JP, *et al.* Dietary fat intake and prevention of cardiovascular disease: systematic review. BMJ 2001; 322(7289): 757-63.
[http://dx.doi.org/10.1136/bmj.322.7289.757] [PMID: 11282859]

[52]    Brull F, Mensink RP, Plat J. Plant sterols: functional lipids in immune function and inflammation? Clin Lipidol 2009; 4(3): 355-65.
[http://dx.doi.org/10.2217/clp.09.26]

[53]    Crews C, Hough P, Godward J, *et al.* Study of the main constituents of some authentic hazelnut oils. J Agric Food Chem 2005; 53(12): 4843-52.
[http://dx.doi.org/10.1021/jf047836w] [PMID: 15941325]

[54]    Racette SB, Lin X, Lefevre M, *et al.* Dose effects of dietary phytosterols on cholesterol metabolism: a controlled feeding study. Am J Clin Nutr 2010; 91(1): 32-8.
[http://dx.doi.org/10.3945/ajcn.2009.28070] [PMID: 19889819]

[55] Jiang Q. Natural forms of vitamin E: metabolism, antioxidant, and anti-inflammatory activities and their role in disease prevention and therapy. Free Radic Biol Med 2014; 72: 76-90.
[http://dx.doi.org/10.1016/j.freeradbiomed.2014.03.035] [PMID: 24704972]

[56] Qureshi AA, Qureshi N, Wright JJ, *et al.* Lowering of serum cholesterol in hypercholesterolemic humans by tocotrienols (palmvitee). Am J Clin Nutr 1991; 53(4): 1021S-6S.
[PMID: 2012010]

[57] Ahsan H, Ahad A, Iqbal J, Siddiqui WA. Pharmacological potential of tocotrienols: a review. Nutr Metab (Lond) 2014; 11(1): 52.
[http://dx.doi.org/10.1186/1743-7075-11-52] [PMID: 25435896]

[58] Amaral JS, Casal S, Alves MR, Seabra RM, Oliveira BP. Tocopherol and tocotrienol content of hazelnut cultivars grown in portugal. J Agric Food Chem 2006; 54(4): 1329-36.
[http://dx.doi.org/10.1021/jf052329f] [PMID: 16478256]

[59] Alasalvar C, Amaral JS, Shahidi F. Functional lipid characteristics of Turkish Tombul hazelnut (Corylus avellana L.). J Agric Food Chem 2006; 54(26): 10177-83.
[http://dx.doi.org/10.1021/jf061702w] [PMID: 17177557]

[60] Kornsteiner M, Wagner K-H, Elmadfa I. Tocopherols and total phenolics in 10 different nut types. Food Chem 2006; 98: 381-7.
[http://dx.doi.org/10.1016/j.foodchem.2005.07.033]

[61] Alasalvar C, Pelvan E. Fat-soluble bioactives in nuts. Eur J Lipid Sci Technol 2011; 113(8): 943-9.
[http://dx.doi.org/10.1002/ejlt.201100066]

[62] Qualified Health Claims: Letter of Enforcement Discretion-Nuts and Coronary Heart Disease, Docket No 02P-0505. Washington, DC, USA: Food and Drug Administration 2003.

[63] Alasalvar C, Shahidi F. Natural antioxidants in tree nuts. Eur J Lipid Sci Technol 2009; 111(11): 1056-62.
[http://dx.doi.org/10.1002/ejlt.200900098]

[64] Dziedzic SZ, Hudson BJ. Polyhydroxy chalcones and flavanones as anti-oxidant for edible oils. Food Chem 1983; 12(3): 205-12.
[http://dx.doi.org/10.1016/0308-8146(83)90007-9]

[65] Andreasen MF, Landbo AK, Christensen LP, Hansen A, Meyer AS. Antioxidant effects of phenolic rye (Secale cereale L.) extracts, monomeric hydroxycinnamates, and ferulic acid dehydrodimers on human low-density lipoproteins. J Agric Food Chem 2001; 49(8): 4090-6.
[http://dx.doi.org/10.1021/jf0101758] [PMID: 11513715]

[66] Alasalvar C, Hoffman AM, Shahidi F. Antioxidant activities and phytochemicals in hazelnut (*Corylus avellana* L.) and hazelnut by-products. In: Alasalvar C, Shahidi F, Eds. Tree nuts Composition, phytochemicals and health effects. Boca Raton: CRC Press 2009; pp. 215-36.

[67] Delgado T, Malheiro R, Pereira JA, Ramalhosa E. Hazelnut (*Corylus avellana* L.) kernels as a source of antioxidants and their potential in relation to other nuts. Ind Crops Prod 2010; 32(3): 621-6.
[http://dx.doi.org/10.1016/j.indcrop.2010.07.019]

[68] Shahidi F, Alasalvar C, Liyana-Pathirana CM. Antioxidant phytochemicals in hazelnut kernel (*Corylus avellana* L.) and hazelnut byproducts. J Agric Food Chem 2007; 55(4): 1212-20.

[http://dx.doi.org/10.1021/jf062472o] [PMID: 17249682]

[69]   Pelvan E, Alasalvar C, Uzman S. Effects of roasting on the antioxidant status and phenolic profiles of commercial Turkish hazelnut varieties (*Corylus avellana L.*). J Agric Food Chem 2012; 60(5): 1218-23.
[http://dx.doi.org/10.1021/jf204893x] [PMID: 22224708]

[70]   Schimitzer V, Slatnar A, Veberic R, Stampar F, Solar A. Roasting affects phenolic composition and antioxidative activity of hazelnut. J Food Sci 2011; 76: 14-9.
[http://dx.doi.org/10.1111/j.1750-3841.2010.01898.x]

[71]   Alasalvar C, Karamać M, Amarowicz R, Shahidi F. Antioxidant and antiradical activities in extracts of hazelnut kernel (*Corylus avellana* L.) and hazelnut green leafy cover. J Agric Food Chem 2006; 54(13): 4826-32.
[http://dx.doi.org/10.1021/jf0601259] [PMID: 16787035]

[72]   Yurttas HC, Schafer HW, Warthesen JJ. Antioxidant activity of nontocopherol hazelnut (*Corylus* spp.) phenolics. J Food Sci 2000; 65(2): 276-80.
[http://dx.doi.org/10.1111/j.1365-2621.2000.tb15993.x]

[73]   Altun M, Celik SE, Guclu K, Ozyurek M, Ercag E, Apak R. Total antioxidant capacity and phenolic contents of turkish hazelnut (*Corylus avellana* L.) kernels and oils. J Food Biochem 2013; 37(1): 53-61.
[http://dx.doi.org/10.1111/j.1745-4514.2011.00599.x]

[74]   Jakopic J, Mikulic-Petkovsek M, Likozar A, Solar A, Stampar F, Veberic R. HPLC-MS identification of phenols in hazelnut (*Corylus avellana* L.) kernels. Food Chem 2011; 124(3): 1100-6.
[http://dx.doi.org/10.1016/j.foodchem.2010.06.011] [PMID: 25212343]

[75]   Amaral JS, Ferreres F, Andrade PB, *et al.* Phenolic profile of hazelnut (*Corylus avellana* L.) leaves cultivars grown in Portugal. Nat Prod Res 2005; 19(2): 157-63.
[http://dx.doi.org/10.1080/14786410410001704778] [PMID: 15715260]

[76]   Amaral JS, Valentao P, Andrade PB, Martins RC, Seabra RM. Phenolic composition of hazelnut leaves: Influence of cultivar, geographical origin and ripening stage. Sci Hortic (Amsterdam) 2010; 126(2): 306-13.
[http://dx.doi.org/10.1016/j.scienta.2010.07.026]

[77]   Del Rio D, Calani L, Dall'Asta M, Brighenti F. Polyphenolic composition of hazelnut skin. J Agric Food Chem 2011; 59(18): 9935-41.
[http://dx.doi.org/10.1021/jf202449z] [PMID: 21819158]

[78]   Nijveldt RJ, van Nood E, van Hoorn DE, Boelens PG, van Norren K, van Leeuwen PA. Flavonoids: a review of probable mechanisms of action and potential applications. Am J Clin Nutr 2001; 74(4): 418-25.
[PMID: 11566638]

[79]   Hertog MG, Kromhout D, Aravanis C, *et al.* Flavonoid intake and long-term risk of coronary heart disease and cancer in the seven countries study. Arch Intern Med 1995; 155(4): 381-6.
[http://dx.doi.org/10.1001/archinte.1995.00430040053006] [PMID: 7848021]

[80]   Wang X, Ouyang YY, Liu J, Zhao G. Flavonoid intake and risk of CVD: a systematic review and

meta-analysis of prospective cohort studies. Br J Nutr 2014; 111(1): 1-11.
[http://dx.doi.org/10.1017/S000711451300278X] [PMID: 23953879]

[81]   Middleton E Jr. Effect of plant flavonoids on immune and inflammatory cell function. Adv Exp Med Biol 1998; 439: 175-82.
[http://dx.doi.org/10.1007/978-1-4615-5335-9_13] [PMID: 9781303]

[82]   Slatnar A, Mikulic-Petkovsek M, Stampar F, Veberic R, Solar A. HPLC-MS$^n$ identification and quantification of phenolic compounds in hazelnut kernels, oil and bagasse pellets. Food Res Int 2014; 64: 783-9.
[http://dx.doi.org/10.1016/j.foodres.2014.08.009]

[83]   Santos-Buelga C, Scalbert A. Proanthocyanidins and tannin-like compounds – nature, occurrence, dietary intake and effects on nutrition and health. J Sci Food Agric 2000; 80: 1094-117.
[http://dx.doi.org/10.1002/(SICI)1097-0010(20000515)80:7<1094::AID-JSFA569>3.0.CO;2-1]

[84]   Williamson G, Manach C. Bioavailability and bioefficacy of polyphenols in humans. II. Review of 93 intervention studies. Am J Clin Nutr 2005; 81(1): 243S-55S.
[PMID: 15640487]

[85]   Kruger MJ, Davies N, Myburgh KH, Lecour S. Proanthocyanidins, anthocyanins and cardiovascular diseases. Food Res Int 2014; 59: 41-52.
[http://dx.doi.org/10.1016/j.foodres.2014.01.046]

[86]   Ouédraogo M, Charles C, Ouédraogo M, Guissou IP, Stévigny C, Duez P. An overview of cancer chemopreventive potential and safety of proanthocyanidins. Nutr Cancer 2011; 63(8): 1163-73.
[http://dx.doi.org/10.1080/01635581.2011.607549] [PMID: 22026415]

[87]   Monagas M, Garrido I, Lebrón-Aguilar R, *et al.* Comparative flavan-3-ol profile and antioxidant capacity of roasted peanut, hazelnut, and almond skins. J Agric Food Chem 2009; 57(22): 10590-9.
[http://dx.doi.org/10.1021/jf901391a] [PMID: 19863084]

[88]   Esatbeyoglu T, Wray V, Winterhalter P. Identification of two novel Prodelphinidin A-type dimers from roasted hazelnut skins ( Corylus avellana L.). J Agric Food Chem 2013; 61(51): 12640-5.
[http://dx.doi.org/10.1021/jf404549w] [PMID: 24313330]

[89]   Gu L, Kelm MA, Hammerstone JF, *et al.* Concentrations of proanthocyanidins in common foods and estimations of normal consumption. J Nutr 2004; 134(3): 613-7.
[PMID: 14988456]

[90]   O'Neil CE, Fulgoni VL III, Nicklas TA. Tree Nut consumption is associated with better adiposity measures and cardiovascular and metabolic syndrome health risk factors in U.S. Adults: NHANES 2005-2010. Nutr J 2015; 14: 64.
[http://dx.doi.org/10.1186/s12937-015-0052-x] [PMID: 26123047]

[91]   Tey SL, Gray AR, Chisholm AW, Delahunty CM, Brown RC. The dose of hazelnuts influences acceptance and diet quality but not inflammatory markers and body composition in overweight and obese individuals. J Nutr 2013; 143(8): 1254-62.
[http://dx.doi.org/10.3945/jn.113.174714] [PMID: 23761651]

[92]   Durak I, Köksal I, Kaçmaz M, Büyükkoçak S, Cimen BM, Oztürk HS. Hazelnut supplementation enhances plasma antioxidant potential and lowers plasma cholesterol levels. Clin Chim Acta 1999;

284(1): 113-5.
[http://dx.doi.org/10.1016/S0009-8981(99)00066-2] [PMID: 10437650]

[93]    Mercanligil SM, Arslan P, Alasalvar C, *et al.* Effects of hazelnut-enriched diet on plasma cholesterol and lipoprotein profiles in hypercholesterolemic adult men. Eur J Clin Nutr 2007; 61(2): 212-20.
[http://dx.doi.org/10.1038/sj.ejcn.1602518] [PMID: 16969381]

[94]    Rajaram S, Sabaté J. Nuts, body weight and insulin resistance. Br J Nutr 2006; 96(1) (Suppl. 2): S79-86.
[http://dx.doi.org/10.1017/BJN20061867] [PMID: 17125537]

[95]    Yücesan FB, Orem A, Kural BV, Orem C, Turan I. Hazelnut consumption decreases the susceptibility of LDL to oxidation, plasma oxidized LDL level and increases the ratio of large/small LDL in normolipidemic healthy subjects. Anadolu Kardiyol Derg 2010; 10(1): 28-35.
[http://dx.doi.org/10.5152/akd.2010.007] [PMID: 20150001]

[96]    Orem A, Yucesan FB, Orem C, *et al.* Hazelnut-enriched diet improves cardiovascular risk biomarkers beyond a lipid-lowering effect in hypercholesterolemic subjects. J Clin Lipidol 2013; 7(2): 123-31.
[http://dx.doi.org/10.1016/j.jacl.2012.10.005] [PMID: 23415431]

[97]    Tey SL, Brown RC, Chisholm AW, Delahunty CM, Gray AR, Williams SM. Effects of different forms of hazelnuts on blood lipids and α-tocopherol concentrations in mildly hypercholesterolemic individuals. Eur J Clin Nutr 2011; 65(1): 117-24.
[http://dx.doi.org/10.1038/ejcn.2010.200] [PMID: 20877394]

# CHAPTER 8

# Bioactive Compounds in Coffee as Health Promotors

Mafalda C. Sarraguça[1], Ricardo N.M.J. Páscoa[1], Miguel Lopo[1], Jorge M.G. Sarraguça[1], João A. Lopes[2,*]

[1] *LAQV/REQUIMTE, Departamento de Ciências Químicas, Faculdade de Farmácia, Universidade do Porto, Porto, Portugal*

[2] *iMed.ULisboa, Departamento de Farmácia Galénica e Tecnologia Farmacêutica, Faculdade de Farmácia, Universidade de Lisboa, Lisbon, Portugal*

**Abstract:** Coffee is the most consumed beverage in the world after water. In 2014 approximately 141 million tons of coffee bags were produced. In terms of international trade only crude oil has a bigger share. The world coffee trade is increasing every year showing the importance of coffee to the world economy. The composition of the two main coffee species (Arabica and Robusta) varies according to the origin, storage and terroir conditions. During the roasting process there are a number of reactions that give rise to the organoleptic properties of coffee. The main bioactive compounds in coffee are chlorogenic acids, caffeine, trigonelline, melanoidins and diterpenes. These compounds are known to have a number of beneficial health effects. Many epidemiological studies suggest that coffee consumption can lead to health benefits in several diseases such as type 2 diabetes, several types of cancers, Parkinson's and Alzheimer's disease. These benefits are related with coffee antioxidant, anti-inflammatory, anti-mutagenic and anti-carcinogenic properties. Chlorogenic acids are known to have chemopreventive and anticarcinogenic activities and also to act as antithrombotic agents. Caffeine is the most recognized bioactive constituent of coffee and can have a number of positive effects in health, most of them associated with the antagonism of the A1 and A2 subtypes of the adenosine receptor. Its stimulatory effect is due to the synergetic interaction with adrenalin and noradrenaline. Trigonelline is connected to neuroprotective, estrogenic, hypoglycemic, anti-invasive, and antibacterial responses.

\* **Corresponding author João A. Lopes:**Departamento de Farmácia Galénica e Tecnologia Farmacêutica, Faculdade de Farmácia, Universidade de Lisboa, Lisbon, Portugal; Tel. +351 217946400; Fax. +351 217946470; Email: jlopes@ff.ulisboa.pt.

The biological activities commonly associated with melanoidins are antioxidant and metal chelating, antimicrobial, and anticarcinogenic. These compounds also have the ability to modulate colonic microflora. Research has showed that the diterpenes, cafestol and kahweol have a chemopreventive potential by enhancing defense systems against oxidative stress. It is clear from the epidemiological studies that coffee has indeed health benefits. Nevertheless some caution has to be taken into account since there are a number of issues regarding these studies, as many of them were not designed specifically for coffee. Furthermore, health problems history and individual lifestyle can introduce misleading factors.

**Keywords:** Alzheimer's disease, Antioxidant, Bioactive compounds, Caffeine, Cancer, Chlorogenic acids, Coffee, Diterpenes, Health benefits, Melanoidins, Parkinson's diseasee, Trigonelline, Type 2 diabetes.

## INTRODUCTION

The coffee plant belongs to the genus *Coffea* [1] and the most important species are *Coffea canephora* (Robusta coffee) and *Coffea Arabica* (Arabica coffee) [2]. Approximately 75% of world coffee production derives from the Arabica coffee species that are considered to have better organoleptic characteristics, while the Robusta coffee species provide the remaining world production and are more resistant to plagues [3]. Coffee is cultivated in tropical areas, Brazil, Vietnam, and Colombia being the main producers and responsible for more than 50% of the worldwide production [4].

There are references in ancient manuscripts dating as far back as 575 AD showing that the Arabs were the first promoters of coffee culture [3] and according to some legends, coffee trees originated in the Ethiopian province of Kaffa [4]. The first known occurrence of coffee beans roasting and conversion into a beverage dates from the XVI century in Persia. Since then coffee dissemination started, and coffee plant arrived in Europe around 1615 AD, brought by travellers. The first cultivation of coffee in Europe was in the botanical garden of Amsterdam, followed by French cultivations in the islands of Sandwich and Bourbon. It took several years before the first coffee house was opened. This happened in the middle of the XVII century in England. Due to a higher demand of the beverage, coffee plantation expanded to European colonies in Africa and South/Central America [3]. It is curious that, until the 20th century, coffee divided the scientific community whether it should be considered as food or medicine due to its long list of human health benefits.

Nowadays, coffee is very much appreciated around the world (mainly in the developed countries) and is the most consumed beverage in the world after water (only crude oil has bigger share in the international trade market) [3]. In 2015, around 141 million tons of coffee bags (60 kg each bag) were reported as the annual worldwide production for 2014 by the International Coffee Organization [4]. The production of coffee is mainly located in developing countries where it plays a crucial role in obtaining foreign exchange earnings as well as tax income and gross domestic product. The annual exports of coffee in 2010 were estimated in US$15.4 billion with about 26 million persons in 52 countries involved [4, 5]. These numbers show the substantial impact of this industry over world's economy and it is expected to continue to grow year after year. The product trade is carried out mainly as green coffee beans. Its price depends on coffee species and variety, geographic location, the methodology used in the processing of green coffee beans and also the care taken during production. The top three importing countries/regions, European Union, USA and Japan, are responsible for more than 70% of total coffee imports in 2015 [4].

The composition of green coffee beans varies according to the species, origin, storage and terroir conditions (the composition of the soil and its fertilisation, altitude, weather). The degree of maturation of the green coffee beans is also critical to obtain high quality coffee. Due to this fact, the harvesting process is only performed when the majority of fruits are ripe through a mechanical or manual process.

The major components of green coffee beans for both Arabica and Robusta species are: carbohydrates, lipids, proteins, chlorogenic acids (CGAs), minerals, trigonelline and caffeine [3]. The most abundant carbohydrate is sucrose which acts as an aroma precursor during roasting [6]. Trigonelline is a pyridine derivative present at high levels that contributes indirectly to the formation of desirable flavour products during coffee roasting [7]. Proteins are also responsible for the development of aroma and taste due to their interaction with other aroma compounds and their role in the formation of melanoidin compounds [8]. The lipids present in coffee prevent the volatilization and loss of flavour during the roasting process [9]. Caffeine is a methylxanthine that contributes to the perceived bitterness of the coffee beverage [10]. Chlorogenic acids are phenolic compounds that have an important contribution in the coffee flavour, coffee cup quality and present several positive health effects [7].

During the roasting process, several interactions and chemical reactions take place between the chemical compounds initially present in green coffee beans. These include a decrease in carbohydrates, chlorogenic acids and the formation of melanoidin compounds. All these changes are responsible for the development of organoleptic properties (aroma, flavour, taste, color) and are, probably, what makes coffee enjoyable [3].

To prepare instant coffee, the green coffee beans need to be roasted, grinded and passed through hot water or steam [3]. Thus, the preparation of instant coffee produces a huge amount of solid residues called spent coffee grounds (SCG). On average, one ton of green coffee generates about of 650 kg of SCG [1]. In the near future, finding eco-friendly alternatives for the disposal of this residue due to its toxic nature will be of the utmost importance.

## BIOACTIVE COMPOUNDS IN COFFEE AND THEIR PHYSIOLOGICAL EFFECTS

From the thousands of compounds present in coffee there are a few that have been considered to have important bioactive properties. The most important ones are: chlorogenic acids (CGAs), caffeine, trigonelline (TRG), melanoidins and the diterpenes (kahweol and cafestol). These compounds are considered to have a positive impact in several diseases due to their antioxidant, anti-carcinogenic, anti-mutagenic and anti-inflammatory properties among others.

### Chlorogenic Acids

Coffee is one of the most important sources of phenolics within a normal human diet, with a contribution of up to 350 mg per 7-oz cup [11]. Phenolics typically exhibit one or more aromatic rings with at least one hydroxyl group attached. So far, over 8000 phenolic structures, present throughout the plant kingdom, have been reported in the literature. These compounds are highly diverse and can range from low molecular weight, simple, single aromatic-ring compounds to large and complex molecules such as tannins and derived polyphenols. The most abundant phenolic compounds found in coffee are chlorogenic acids (CGAs). These compounds can make up to 12% of the whole dry matter of green coffee beans [12]. In the coffee plant, CGAs are formed through esterification of trans-cinnamic acids, particularly caffeic, ferulic and p-coumeric, with quinic acid. The most abundant CGAs in coffee are caffeic acid including 5-O-caffeoylquinic (5-CQA), which accounts for 50% of the total chlorogenic acid content, (Fig. **1**) and its isomers 3- and 4-CQA [13]. Significant amounts of the three analogous feruloylquinic acids as well as 3,4-O-, 3,5-O- and 4,5-O-dicaffeoylquinic acids and 3-O- and 4-O-caffeoylquinic acid are also present [14]. More recently, several mono-acyl and diacyl CGAs including p-coumaric acid and 3,4-dimethox--cinnamic acid have also been characterised in green coffee beans [15]. CGAs are present in high concentration in coffee beans and are responsible for coffee flavor as well as its quality. Despite the fact that green coffee beans total content of CGAs is mainly dependent on genetics (*i.e.* species), the degree of maturation, agricultural practices as well as climate and soil of the plantation also play an important role [16]. Studies have revealed that human intervention (mainly through processing of the coffee beans) has a significant effect on the CGAs profile of green coffee. During the roasting process, the loss of a water molecule from the quinic acid moiety forms an intramolecular ester bond, leading to the formation of chlorogenic acid lactones. The high temperatures occurring during coffee roasting thus cause a reduction in total CGAs, which seems to be proportional to the intensity of roasting [17].

However, the chemical transformations that occur to CGAs during the roasting process are yet to be completely unveiled. The final content of CGAs in the coffee beverage is, therefore, dependent on several factors, among which coffee species, bean variety and processing conditions are some of the most important.

CGAs have been refered as powerful antioxidants. However, after ingestion of coffee, unmetabolised CGAs achieve a very low concentration in the plasma. The health benefits that may come from a direct and simplistic antioxidant hypothesis have repeatedly been challenged and have now been discarded [18]. Studies have reported CGAs to be able to modulate the expression of specific genes encoding enzymes involved in phase II metabolism that play an important part in the endogenous antioxidant defenses, thus acting as chemopreventive agents [19]. There are several other mechanisms through which CGAs might exert anticarcinogenic activities: the inhibition of deoxyribonucleic acid (DNA) methyltransferase being one example. This process plays an important role in cell differentiation and cellular aging among others [20]. It may also promote inhibition of platelet activity and therefore work as an antithrombotic agent [21].

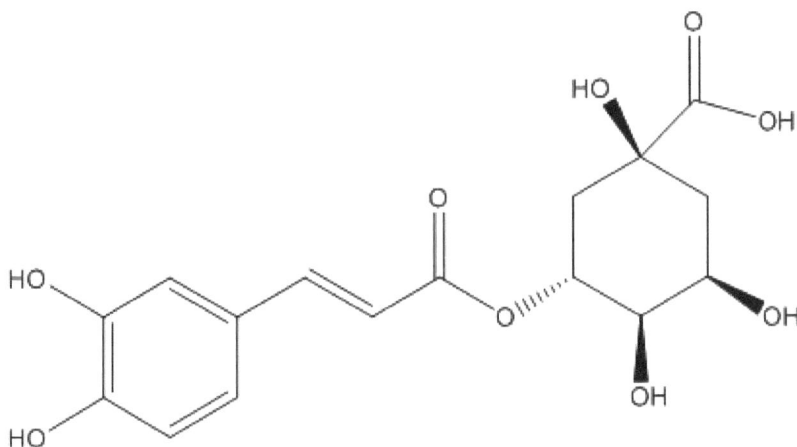

**Fig. (1).** Chemical structure of 5-*O*-caffeylquinic acid (5-CQA).

It is evident from *in vivo* and *ex vivo* studies that coffee has a protective effect, and it is important to use CGA-derivatives present in the circulatory system for the assessment of potential health benefits and to gain a deeper understanding of the mechanism of action of these compounds [18].

## Caffeine

The alkaloid caffeine is probably the most recognized and prominent bioactive constituent from coffee and is among the most commonly consumed stimulants worldwide. It is characterized by its bitter taste and white crystalline powdery aspect. Caffeine (1,3,7-trimethylxanthine) (Fig. **2**) was first isolated from coffee in 1820 [22, 23], it contains two fused rings that are known to be related to purines. Coffee is the richest known source of caffeine (*e.g.* 240-mL of instant coffee comprises circa 100 mg of caffeine) [24]. Caffeine content is typically found at levels between 0.5-3.5% in the coffee bean [25], depending on several factors such as variety, environmental and agricultural factors, as well as processing techniques. Caffeine content of brewed coffee varies greatly and is mainly dependent on the specific preparation method as well as the raw material itself. Be that as it may, average levels of caffeine in coffee vary typically between 50 and 300 mg per 8-oz serving. One study has described a wide fluctuation in caffeine concentrations in coffee beverages [259-564 mg/dose] obtained from the same commercial establishment on six consecutive days [26]. Roasting also affects the caffeine content of coffee, more intense roasting leading to reduced caffeine content of the beans [27].

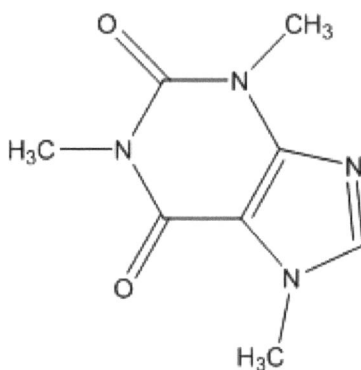

**Fig. (2).** Chemical structure of caffeine.

Caffeine is very quickly and almost completely absorbed in the stomach and small intestine, and it is consequently distributed to all tissues. After absorption, it exhibits numerous physiological effects, exerting most of them through antagonisation of both the A1 and A2 subtypes of the adenosine receptor [18, 28]. This process leads to a rise of dopamine levels within the organism. Dopamine is responsible for many of the central nervous system stimulating properties of caffeine. It also interacts with adrenalin and noradrenaline which are the main neurotransmitters for the sympathetic nervous system. This synergetic interaction leads to the stimulatory effect of caffeine. Caffeine also has a positive effect in long-term retention memory consolidation [29]. It helps to reduce symptoms linked to Parkinson's disease (PD) due to improvements in the performance of the dopamine synergetic system. By blocking A2 adenosine receptors, caffeine stimulates the release of dopamine [30]. Moreover, caffeine increases the metabolic rate leading to an important help in weight loss and decreasing the risk of several diseases associated with metabolic syndrome [28].

The large number of health benefits has a counterpart in negative health impacts. The balance between them is difficult to define since it depends largely on the individual susceptibility to caffeine as well as other factors [18].

## Trigonelline

Trigonelline (TRG) (N-methyl nicotinic acid) is a nitrogenous compound, a pyridine alkaloid derived from the methylation of the nitrogen atom of nicotinic acid (niacin), hence the name N-methyl nicotinic acid [31] (Fig. **3**). TRG is the second main alkaloid (after caffeine) present in coffee beans. During roasting, TRG undergoes extreme thermal degradation, generating several volatile compounds. Some of these compounds (*e.g.* the pyridine and pyrrole derivatives), are important to the coffee beverage flavour [32]. Furthermore, due to the acidic conditions originated during roasting, TRG is demethylated originating niacin, a water soluble vitamin B complex also known as nicotinic acid [33]. TRG is normally used to assess and define roasting intensity in both Arabica and Robusta coffees [34].

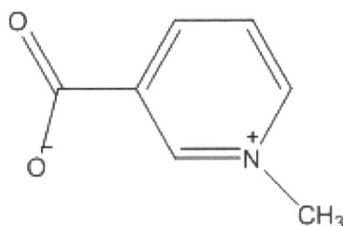

**Fig. (3).** Chemical structure of trigonelline.

After ingestion of coffee, TRG is mainly absorbed in the stomach. Regarding its biological activities, studies have revealed that this compound has antibacterial, anti-invasive, estrogenic, neuroprotective and hypoglycemic activities [18]. Research with diabetic rats showed that after treatment with TRG the levels of triglycerides, total cholesterol and blood glucose decreased [35].

In a study with healthy men, a dosage of this compound led to a reduction in insulin levels, after ingestion of glucose, when compared with a placebo [36]. Some studies with rodent models of Parkinson's disease also showed a neuroprotective effect of TRG [37]. Anti-invasive activity of TRG was detected in studies with rat cells. From these results it was possible to infer that TRG and TRG-loaded rat serum inhibited reactive oxygen species (ROS). This suggests a mechanism different from a direct antioxidant effect [38]. The estrogenic activity attributed to TRG is due to the activation of the estrogen receptor which classifies this compound as a phytoestrogen [39]. It has been reported that TRG increases the sensitivity to anticancer drugs of chemoresistant pancreatic and colon cell lines, acting as an effective inhibitor of the Nrf2, a transcription factor that plays an important role in cancer development [40]. This compound also has impacts on the adhesive properties of bacteria causing caries in humans, and is also a potential microbial agent against *Salmonella enterica* [41].

**Melanoidins**

Melanoidins (Fig. **4**) are nitrogen-containing, heterogeneous brown pigments produced by the Maillard reaction (MR) or caramelization of carbohydrates which occurs during the roasting process of the coffee beans [42]. In coffee, the MR usually takes place within reaction times of less than 2 h, at temperatures higher than 150 °C and lower moisture environments. Melanoidins are normally generated from the reduction of sugars and proteins or amino acids in the late stages of the MR. They are largely responsible for the characteristic brown color of coffee and are able to absorb light at wavelengths up to 420 nm. The complexity of melanoidins produced in the MR depends mostly on the source origin and the technological conditions of the reaction (*e.g.* temperature, time, pH, solvent, *etc.*) [43]. Melanoidins are formed by condensations, isomerizations, rearrangements, dehydrations, retroaldolizations and cyclizations of MR products with low molecular weight. However, due to the complexity of the products that are generated in the reaction, the chemical composition of melanoidin structures is fairly unknown [44]. Melanoidins found in coffee are negatively charged but are heterogenous when it comes to their polyanionic behaviour. High molecular weight (HMW) melanoidins of coffee are known to be more negatively charged than low molecular weight (LMW) ones [45]. Researchers have hypothesized that CGAs could be the cause of negative charges in coffee melanoidins. Even though melanoidins are considered HMW compounds, more recent results have indicated that they also have a LMW fraction [43]. Long roasting periods lead mostly to the formation of HMW melanoidins which display a more profound brown color than LMW ones [46]. More than half (59%) of the melanoidins in coffee are HMW (>12–14 kDa). Overall, little is known about melanoidins structural properties, even though numerous attempts have been made to isolate and purify these compounds [47].

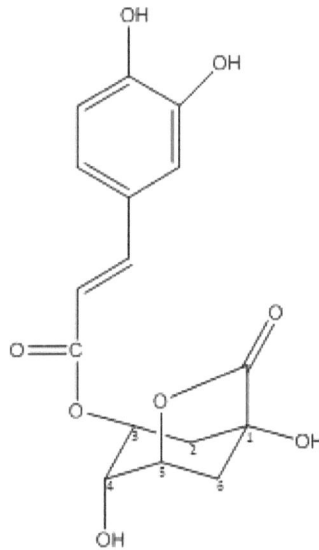

**Fig. (4).** Chemical structure of 3-caffeoylquinic-1,5-γ-lactone.

Because coffee melanoidins are almost indigestible by humans, they tend to be fermented in the alimentary canal and act, therefore, as dietary fibers. It has been estimated that 0.5 to 2.0 g of melanoidins per day reaches the colon of an individual with a moderate/heavy coffee ingestion. This contributes significantly to health benefits at the colon level [48]. Different activities with putative biological benefits have been associated to these compounds, *e.g.*, antimicrobial and anticarcinogenic activity as well as antioxidant and metal chelating activity. Melanoidins have also been reported to have an antihypertensive and antiglycative activity. The ability to modulate colonic microflora is another characteristic of these compounds [18].

## Diterpenes: Kahweol and Cafestol

Diterpenes are a group of terpenoids commonly found as secondary metabolites in terrestrial and marine organisms. They often occur in a C-20 backbone where isoprene units combine in different forms to give an array of diterpenes such as cafestol (Fig. **5a**) and kahweol (Fig. **5b**) [49]. Levels of diterpenes in coffee beans depend on the coffee species and storage conditions as kahweol and cafestol are sensitive to acids, heat and light [50]. Kahweol and cafestol are chemicals only found in coffee beans and brews. They are present in the unsaponifiable fraction of the lipid phase of coffee beans and/or grounds and are highly unstable molecules that easily form oxides [51]. Coffee beans contain 0.27-0.67% (Arabica) and 0.15-0.37% (Robusta) dry matter of cafestol. Similarly kahweol content is 0.11-0.35% and 0.1% dry matter in Arabica and Robusta beans, respectively [52]. Roasting seems to have an influence on diterpenes profiles of coffee beans. High roasting temperatures coupled with prolonged roasting times lead to the degradation of both of these compounds to dehydrokahweol and dehydrocafestol respectively, after 8 min of roasting process [53].

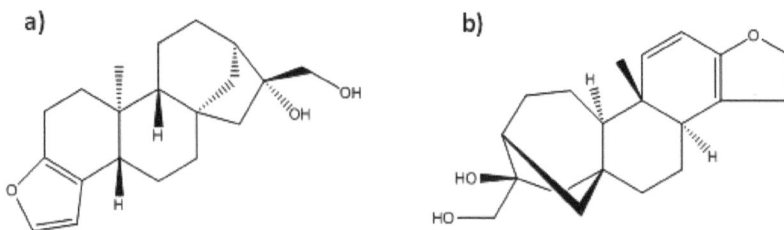

**Fig. (5).** Chemical structure of diterpenes: **a)** cafestol and **b)** kahweol.

It has been shown that diterpenes have a chemopreventive potential [54]. This is achieved mainly through the enhancement of defence systems against oxidative stress, the regulation of Nrf2/ARE signaling pathways and the induction of phase II detoxifying enzymes [55].

## HEALTH BENEFITS OF COFFEE

Many epidemiological studies suggest that coffee consumption can lead to health benefits in several diseases such as type 2 diabetes (T2D), several cancers, Parkinson's and Alzheimer's diseases (AD). The health benefits of coffee have been attributed to its antioxidant, anti-inflammatory, anti-mutagenic and anti-carcinogenic activities [49]. However, there are a number of problems associated with these studies, as many were not designed specifically for coffee. Furthermore, other health issues and personal attributes can introduce confounding factors. Even with the problems associated with the epidemiological studies there is an increasing number of studies that lead to the conclusion that coffee has a beneficial effects in human health.

### Antioxidant Effect

Green coffee beans are rich in CGAs that are usually referred as powerful antioxidants [56]. The roasting process affects the composition of the CGAs present in coffee leading to the generation of different compounds deriving from the MR, carbohydrate caramelization and pyrolysis of organic compounds [57]. Due to the new compounds produced during the roasting process the antioxidant activity of roasted coffee increases when compared with green coffee.

Several studies analyzed the effect of coffee consumption on the antioxidant activity in the plasma. The increase in glutathione-S-transferase activity and glutathione has been suggested to be beneficial to health [58, 59]. After a daily consumption of coffee for two weeks, a 15% increase in glutathione in the plasma was observed [58]. Similar results were found after a consumption of five cups a day during one week [59].

Coffee consumption can contribute to the endogenous system which prevents oxidative damage of DNA, cell components, proteins and lipids which help prevent several type of diseases such as degenerative and neurodegenerative diseases, cancer and cardiovascular diseases.

## Diabetes

A large number of prospective studies and meta-analysis studies indicate that a moderated coffee consumption reduces the risk of T2D [60]. The association between coffee intake and the reduction of T2D risk is independent of race [61], gender [62, 63] and geographic distribution of the population. There are studies in Europe [62 - 66] in America [61, 67 - 69] and Asia [70 - 72].

A prospective study involving 17,000 Dutch men and women reported a decrease in 50% in the risk of developing T2D when 7 cups of coffee a day are consumed compared with those who had an intake lower than 2 cups per day [62]. A similar result was obtained in a study involving Swedish women with a coffee consumption of three or more cups a day in relation with two or less cups per day [73]. A meta-analysis with 15 epidemiological studies, 9 prospective and 6 case-control involving a total of 200,000 participants showed that with an intake of four or more cups a day the reduction of the risk of T2D was 35% [74]. In a second meta-analysis involving 500,000 people with follow-up periods ranging from 2 to 20 years from 20 prospective studies indicated that the inverse association between coffee intake and the risk of T2D was dosage dependent with a decrease of 7% per cup of coffee consumed per day. The association was also valid for decaffeinated coffee. In this case the analysis focused on 6 studies with a total of 225,000 people with a 33% decrease in the risk of T2D [75]. Coffee consumption is also constantly associated with a lower occurrence of impaired glucose intolerance, hyperinsulinemia, hyperglycemia and insulin sensitivity [60].

## *Mechanism of Action*

Many of the compounds present in coffee affect the progress of diabetes due to the role played in glucose metabolism. There are several studies proposing that the phenolic compounds can reduce fasting plasma glucose, increasing sensitivity to insulin. It is also known that several metabolites of CGAs also exert a hypoglycemic effect. Trigonelline and magnesium were also suggested to have an effect in glucose metabolism [60]. Some authors hypothesize that the protective effect of coffee could be due to the thermogenic effect of caffeine that indirectly reduce the risk of obesity [76]. Chronic hyperglycemia which is induced by oxidative stress has a significant part in the mechanism leading to pancreas beta-cells dysfunction [77]. The pancreas beta-cells could be protected from oxidative stress by the antioxidants present in the coffee [78]. Subclinical inflammation is deeply involved in the pathogenesis of T2D. This is why the protective effect of coffee regarding the inflammation process due to the phenolic compounds may reduce the risk of T2D [60]. Caffeine as also been proposed as having an anti-inflammatory effect which, at high dosages, could protect pancreas beta-cells [79].

## Cancer

Currently there are some evidences for the inverse association between coffee consumption and the risk of several types of cancer such as breast and uterus cancer, prostate, liver and colorectal.

### Breast Cancer

Until 2007 only three studies were available and were considered by the World Cancer Research Fund insufficient to draw any conclusions between coffee intake and breast cancer [80].

From 2009 until now four meta-analysis were published [81 - 84]. In all of them an inverse association between coffee intake and breast cancer was found. Another case-control found that a significant inverse association could be found with a coffee consumption of five or more coffees per day [85]. It was also found that coffee can prevent early events in tamoxifen-treated breast cancers patients and modulates hormone receptor status [86].

### Uterus Cancer

Three control studies on the relation between coffee consumption and uterus cancer indicated a reduction of 50-60% in the risk with an intake of three or more cups of coffee a day [87 - 89]. Similar results were found in a cohort study with fifteen years of follow-up in Japan [90].

## Prostate Cancer

Until now no clear evidence between overall prostate cancer risk and coffee intake was found [55]. However, for the case of lethal and advanced prostate cancer an inverse association between coffee consumption and the prostate cancer was found in two different studies [91, 92]. An up-to-date study reported an association between a large coffee intake and a reduced risk of prostate cancer recurrence/progression [93].

## Liver Cancer

Since 2007 four meta-analysis studies in which the relation between coffee intake and the risk of liver cancer was analyzed concluded that there is an inverse association between coffee consumption and liver cancers [83, 94 - 96]. In 2013 a cohort study involving 27,037 Finnish male smokers determined that coffee intake led to a decrease in incident liver cancer and in mortality from chronic liver disease [97].

It has been pointed out that subjects with liver conditions may selectively reduce the coffee intake [96]. However, there is a consistency across the studies and geographic areas that support the claim that coffee consumption has an inverse relation with liver cancer [18].

## Colorectal Cancer

There have been a large number of studies regarding the association between colorectal cancer and coffee consumption. An evaluation of 15 control and 3 cohort studies lead to the conclusion that there is an opposite association between the risk of colorectal cancer and coffee intake for some control-studies but the results were inconsistent in the cohort studies [98]. Two meta-analysis, one comprising 20 control-cases studies and the second with 25 control-cases and 16 cohort studies, revealed a moderated inverse relation between coffee and colorectal cancer [99, 100]. Another study confirmed these results and concluded that a dosage of 4 cups per day had an inverse association with colorectal cancer [101].

Two Japanese studies observed that a significant decrease in the risk of colorectal cancer was associated with a moderate intake of coffee (3 cups a day) [102, 103].

The accumulating evidence suggests a modest protective effect of coffee in the risk of colorectal cancer [18, 55].

## Mechanisms of Action

There are several factors known to be connected with cancer risk, such as obesity, lack of physical activity, alcohol consumption, diet, among other stress related reasons [104]. Infections are one of the typical examples of stressors that produce increased amounts of reactive oxygen species (ROS). The transformation of a healthy cell into a cancerous cell is a multistep procedure with the first step being the DNA damage provoked by the ROS. Throughout the transformation, cells develop defects in terminal differentiation, apoptosis and growth control. The final stages involve angiogenesis and finally loss of tissue restrictions and metastasis. Data retrieved form experimental work leads to the conclusion that coffee can reverse or interfere with the different steps of the cancerous process [55].

Coffee is a natural source of antioxidants and is one of the major contributors of antioxidants in the human diet. Therefore, there is the possibility that coffee can reduce ROS by a direct antioxidant effect [105, 106]. The optimization of defense processes which involve endogenous antioxidants, detoxification processes, DNA repair mechanisms and apoptosis is possible by the protective action of coffee against damage by ROS and cancer development [55]. Inflammation is considered an important factor in several diseases including cancer, since chronical inflammation and oxidative damage increase the susceptibility for malignant cells. The ability to modulate immune processes and damped inflammation has been proposed as a mechanism by which coffee may prevent or delay cancer development [107]. Chlorogenic and caffeic acids, caffeine and diterpenes are known to inhibit induced NF- kB activation which is a crucial transcription factor involved in immune and inflammatory processes and over expressed in several cancers [108]. Apoptosis is vital to ensure damaged cells removal. *In vitro* studies show that caffeine [109], cafestol [110] and kahweol [111] can induce apoptosis. Also, caffeine has a protective effect on the UVB-induced carcinogenesis through induced apoptosis and increased caspase 3 activity in murine models of skin cancer [112].

## Parkinson's Disease

A large study comprised of eight case-control and five cohort studies made between 1968 and 2001 in different countries lead to the conclusion that people that drink coffee have a lower risk [31%] of developing Parkinson's disease (PD) [113]. In these studies, conducted separately for men and women, a strong inverse linear relation between the number of cups per day consumed and the risk of PD was found for men, while for women, no relation between the number of cups per day and the risk of PD was observed. Men drinking several cups of coffee had a 49% lower risk of developing the disease. The reason for the difference between men and women is probably due to hormonal effects [114]. These results were corroborated by a cohort study that showed a significant inverse relation between coffee intake and the risk of PD's mortality in men. However, in women this association was dependent on postmenstrual estrogen use [115].

### *Mechanism of Action*

Caffeine is considered to be the reason for the protective effect of coffee in PD. PD is a neuropathological disorder involving the degeneration of dopaminergic neurons in the substantia nigra, with the subsequent loss of their terminals in the striatum [18]. It has been demonstrated that caffeine has beneficial effects regarding dopaminergic neurotoxicity in animal models [116]. This effect may be due to caffeine's ability to block the adenosine A2A receptor [56]. In fact, a prospective study covering 30 years concluded that the intake of decaffeinated coffee did not lower the risk [117].

## Age-Related Cognitive Decline and Alzheimer's Disease

### *Age-related Cognitive Decline*

Normal aging is associated with modest decrease in several aspects of cognitive function, including memory and information processing speed [118].

Six prospective studies reported protective effects in some adults aged 65 or more.

The consumption of coffee or caffeine was about 3 cups per day [117, 119 - 123]. However, it is still difficult to prove that caffeine has a protective effect. Two different studies reported no association between decaffeinated coffee and potential cognitive protection [123, 124].

### Alzheimer's Disease and Dementia

Two prospective studies have associated daily coffee intake with a decrease in AD [125, 126]. A 30% decrease in the risk of AD in adults over 65 years of age was associated with daily coffee intake [126]. In the other study a moderate coffee consumption of 3-5 cups per day of coffee during mid-life was correlated with a reduction of 65% in the risk of AD or late life dementia [125]. A small case-control study of older adults with probable AD that consumed coffee in the 20 years prior to the diagnosis showed a strong inverse correlation (60%) of caffeine intake (2-3 cups) and AD risk [127]. A systematic review and meta-analysis [128] concluded that the existent epidemiological studies show that caffeine has a positive effect in lowering the risk of AD and dementia; however, there are a number of factors that lead to a difficult interpretation of the overall data.

### Mechanism of Action

The neuroprotective potential of caffeine in neurodegenerative pathology has been demonstrated in several AD mouse models. At physiological concentrations, caffeine acts as a non-selective adenosine receptor antagonist. By reducing signaling through adenosine receptors, it modulates the activities of many neurotransmitter systems [118]. CGAs can also have a positive effect in dementia and AD since they benefit biological pathways related to risk of cognitive decline and dementia including hypertension, inflammatory signaling, insulin signaling and glucose metabolism [118].

## ANALYTICAL METHODS FOR THE DETERMINATION OF BIOACTIVE COMPOUNDS IN COFFEE

The bioactive components of coffee have a representation in diverse aspects of the beverage as a product for human consumption. Their direct relation to health is of the greatest importance, with the antioxidant properties taking a special place in this rank. Some bioactive components have implications in the sensory aspects such as bitterness, acidity and flavor and are directly evaluated by the consumer.

These aspects relate directly to the need for quality control through accurate, precise and sensitive analytical methods. Separation-based analytical methods such as gas (GC) and high-performance liquid chromatography (HPLC), mainly coupled with ultraviolet/ visible (UV-VIS) spectroscopy and mass spectrometry (MS) detection, and capillary electrophoresis (CE) take a central part on the analysis of organic compounds. Non-separative techniques have also been used for the analysis of the composition of coffee, namely electrochemical and spectroscopic techniques such as near-infrared spectroscopy (NIRS). Comprehensive reviews of these methods can be found in literature [51, 129]. The adequacy of the method to be used is determined by the characteristics of the component to be determined (*e.g.* volatility and polarity).

HPLC due to its low limits of detection (LOD), versatility and accessibility is known to be central in almost all areas where chemical analysis of composition is needed, and the analysis of coffee composition is not an exception. GC, on the other hand, while having also a low LOD is especially suited for the analysis of more volatile components. CE promotes electrokinetic migration of the analytes to the positive or negative poles by using high voltage and is adequate for the analysis of components with non-neutral net charge. In a study focusing on the determination of caffeine in decaffeinated coffee, it was shown that chemical electrophoresis may be a good alternative to HPLC, presenting lower costs in terms of reagents and a faster time of analysis when compared to a HPLC method [130].

Electrochemical techniques such as voltammetry and coulometry are fast, simple, selective and sensitive techniques in which the oxidation-reduction potential of the compounds is used, exploiting their specific relations with potential and current. These methods are frequently regarded as low cost, fast response and simple to operate. Coulometric titration has been proved useful in studying the mechanisms by which the presence of additives in coffee beverages, such as milk proteins, reduce the antioxidant properties of coffee [131]. Direct determination of major components such as CGAs and caffeine has been achieved with voltammetry [132 - 134].

NIRS has recently been introduced in the study of coffee composition. In this spectroscopic technique, the information bands are created by the combination and overtones of the fundamental infrared frequencies. The functional groups contributing to the information in this spectral region are mainly those in which a hydrogen atom is present (C-H, N-H, O-H and S-H). This technique although not selective and presenting a high LOD, presents several important characteristics that make it interesting for the study of coffee composition. NIRS is a non-destructive, non-invasive technique that requires minimal or no sample preparation. It is a fast and robust technique, with a very low cost per analysis, and can be used in laboratory, field, and in industrial environments. This technique while coupled with multivariate analysis allows not only for a direct assessment of the chemical composition of the samples but also, as it will be demonstrated bellow, for the evaluation of its relation with non-chemical attributes such as sensory and quality attributes that may be of importance.

## Determination of Chlorogenic Acids /Phenolic Compounds

As mentioned in the previous sections, phenolic compounds (mainly CGAs and derivatives) are one of the most important groups of bioactive compounds present in coffee and are generally present in high concentrations. Their presence influences significantly the quality and flavor (aroma and astringency) of coffee. Therefore, it is not surprising that an extensive amount of analytical methods have been developed for the determination of this class of compounds in coffee samples.

Chromatographic methods have been widely used for the identification and quantification of phenolic compounds, with special emphasis on CGAs. The extraction procedure is frequently based on an extraction with 70% (v/v) methanol at 4°C. This method usually includes a sample clean-up step following the evaporation of methanol. This may consist on treatment with Carrez reagent following evaporation of methanol [135], or in the presence of methanol [136], or using lead acetate [137] just to mention a few possibilities. The possibility of Carrez reagent products to react with CGAs [138] prompted the interest in developing alternative methods which led to the application of microwave assisted extraction and water extraction under high-pressure and high-temperature conditions [139, 140].

HPLC-MS has been widely used for studying complex mixtures. Rodrigues *et al.* [137] ,using HPLC–DAD–MS (Diode array detection - MS detection) were able to identify and quantify, in a study comprising brews of 10 roasted ground coffee and 4 soluble coffee, caffeine, trigonelline theobromine, theophylline, 5 chlorogenic acid lactones, 17 chlorogenic acids, 2 cinnamoyl-amino acid conjugates, nicotinic acid, 2 free cinnamic acids, and 5-hydroxymethylfurfural. The presence of caffeoylferuloylquinic acid isomers, and cinnamoyl-amino acid conjugates in soluble coffee brews was reported for the first time in this study. The same study concluded that decaffeinated coffee brews contained higher concentrations of caffeic acid, trigonelline, CGAs, chlorogenic acid lactones, when compared to regular ones. Caffeine was monitored and quantified at 272 nm, p-coumaric acid, at 310 nm, 5-caffeoylquinic acid, ferulic acid and caffeic acid, at 325 nm. When no commercial standards were available, compounds were identified based on the elution order on reverse phase and characteristics of UV–VIS and mass spectra as compared to the literature.

Campa *et al.* developed an HPLC-UV method to access the relation between caffeine and chlorogenic acid contents [135]. In the same paper a description of phenolic compounds profiles by coffee species was presented by quantifying 3- and 4- and 5-caffeoylquinic acid (CQA), 3-, 4-, and 5-feruloylquinic acids (FQA), and 3,4-, 3,5- and 4,5-dicaffeoylquinic at room temperature, with detection set at 325 nm.

Upon roasting, fractions of the CGAs present in the green beans may react by isomerization, dehydration, hydrolization and degradation or reduction [141, 142]. The influence of the roasting process over the CGAs composition of Arabica and Robusta coffees was addressed by Vignoli *et al.* using HPLC-UV [143].

Perrone *et al.* studied the phenolic composition of coffee in green and roasted beans, using LC-MS [144]. For the first time, 1-feruloylquinic lactone, 1-feruloylquinic acid, and 3,4-diferuloylquinic acid were quantified in *C. arabica* and *C. canephora*, adding to nineteen previously identified CGAs and chlorogenic acid lactones, while differentiator compounds were also reported (3- and 4-pcoumaroylquinic lactones were reported in *C. canephora* and 3,4-di-p-coumaroylquinic acid was identified in *C. Arabica*).

The effect of roasting on the concentration of phenolic components was studied by HPLC-PDA (photodiode array detector) [27], (detection at 278 nm) and PDA detection recording the absorbance of the eluate between 200 and 400 nm. The effect of different brewing methods on the phenolic content of coffee was also addressed using HPLC-PDA [145], focusing on content of polyphenols and antioxidant capacity of these brews. Effect of milk addition was also addressed and observed to significantly decrease the content of chlorogenic acid derivates, total phenols, caffeine, and antioxidant capacity.

Regarding non-separative methods, electroanalytical methods are the more frequently used for basic analyses regarding chlorogenic acid quantification. Square wave voltammetry, differential pulse voltammetry, adsorptive stripping voltammetry, based on gold electrode [134], boron doped diamond electrode [132] and carbon paste electrode [146] are some examples.

## Determination of Caffeine

As with phenolic compounds, separation based methods (mainly HPLC) have been commonly used for the determination and quantification of caffeine. Sample preparation for these analyses frequently involves only filtration and dilution [27, 145, 147]. As with phenolic compounds, clarification can be obtained by methods using Carrez reagents [137, 148] or lead acetate [6]. Solid phase extraction [149] and water extraction in the presence of magnesium oxide [12] have also been used for solution clarification.

Caffeine concentration has been frequently determined simultaneously with CGAs by HPLC-UV [145, 148, 150] at different wavelengths from 265 to 278 nm [27, 33, 147, 151]. The caffeine concentration dependence on the species was addressed by Hecimovic *et al.* [2011] by HPLC-UV, while studying the effect of roasting [27]. HPLC-VWD (variable wavelength detector) was used for studying the brewing effects on caffeine concentration [147]. Setting the detector at 270 nm, the combination of different water temperatures and pressures in an espresso coffee (EC) machine was tested to observe the influence on the extraction of Arabica and Robusta coffees in 20 commercial samples.

Analytical methods based on direct spectroscopic analysis have also been proved useful for the determination of caffeine. Direct UV-VIS spectroscopy has been used and compared with the HPLC-UV method [27]. In this work, two different procedures, producing lower results than the reference HPLC method, were tested: one using lead acetate, and the other with benzene and back extraction with $H_2SO_4$. Extraction with dichloromethane has also been used when proceeding in a direct determination by UV spectroscopy for removing matrix effects [24], and monitoring absorbance at 274 nm. Fourier transform infrared spectroscopy coupled with attenuated total reflectance (FTIR-ATR) was observed to provide a fast and relatively simple way of determination [152]. Caffeine concentration was monitored at 1655 $cm^{-1}$. Sample preparation for this analysis involved liquid extraction with chloroform which removes interference from the solvent in the 1650-1700 $cm^{-1}$ range.

NIRS was recently introduced as a viable technique for the determination of caffeine in ground coffee samples [153]. Ground coffee samples covering an extensive range of roasted levels were analyzed by NIRS, from which partial least squares (PLS) prediction models for the caffeine concentration were developed with a calibration based on HPLC-UV method as reference values. The root-mean-square error of prediction (RMSEP) of 0.378 mg/g was determined for the model by using an independent test set, while taking a short analysis time and having the non-destructive advantages of the technique, important for on-line monitoring. NIRS in diffuse reflectance mode was also used for discrimination between coffee varieties of the lyophilized extract. It was concluded that the differentiation was due to caffeine and other alkaloids [154]. NIRS in reflectance mode was also found to be a good alternative to HPLC for the quantification of caffeine and theobromide [155]. Additionally, multivariate analysis based on NIRS was used to correlate aspects of chemical composition with acidity, bitterness, flavor cleanliness body and overall quality. Ribeiro *et al.* [2011] observed that bitterness, acidity, flavor, CGAs, trigonelline, polysaccharides, sucrose, protein content, cleanliness and overall quality were somehow all related to caffeine concentration [156].

Caffeine was also been determined by means of voltammetry techniques, which provides a fast, inexpensive and highly selective solution [133, 157]. When comparing square wave voltammetry and cyclic voltammetry, using a pseudo-carbon paste electrode, the former produced better results illustrated by the formation of a sound oxidation peak [133].

## Determination of Diterpenes

Cafestol and kahweol are two compounds that are unique to coffee. The preferred method for the determination of diterpenes, due to their volatility has been gas chromatography GC-FID (flame ionization detector), although HPLC is also a common alternative technique. Pre-treatment of samples for analysis frequently involves extraction of diterpenes from coffee by Soxhlet extraction, saponification and silylation [158]. Alternative procedures have been proposed, such as transesterification of the esters of diterpenes [159]. The relation between the integrated peak areas of kahweol and 16-O-methyl cafestol was used for authentication purposes [158, 160].

Cafestol and kahweol have been determined by HPLC-UV at 230 nm and 290 nm respectively [161, 151] using SPE and saponification for sample preparation. In comparing different brews, diterpene levels were found to be larger in espresso, French press and Turkish style than in instant or filtered coffee [161]. Kolling-Speer *et al.* [1995] proposed a method for the determination of cafestol and kahweol using a single wavelength (220 nm) for both components [162]. This methodology has been used with a green chemistry approach for the extraction of diterpenes using supercritical carbon dioxide before sample saponification [163] for the study of diterpene levels on green and roasted coffee oil. Oil content levels and diterpene oil concentration were compared with the results obtained by extraction with the Soxhlet apparatus, using hexane as solvent.

The presence of diterpenes and derivative compounds resulting from roasting were determined by liquid chromatography with UV–VIS and MS detection. Dehydro derivatives were found in the darker roast samples and were observed to be dependent on the intensity of the roasting process, as well as the level of diterpenes in both Arabica and Robusta [53].

Isolation and purification of different diterpenes from both Arabica and Robusta varieties has been achieved by high-speed countercurrent chromatography [164] using hexane-ethyl acetate-ethanol-water mixtures as solvent systems.

## Determination of Trigonelline and Nicotinic Acid

Trigonelline and nicotinic acid determination have frequently been addressed in analytical methods literature, together with the determination of caffeine in coffee samples, mainly by HPLC-UV, making the methodology largely similar, with the detail that trigonelline is detected at 265 nm [33] or 260 nm [151]. The influence of brewing in trigonelline and nicotinic acid concentration has also been studied using HPLC-VWD [147]. Decaffeination was reported to decrease trigonelline in percentage when compared to regular coffee, while using HPLC-UV [165].

GC-MS has also been employed to quantify these components and determine products of degradation upon roasting [166]. Also, using LC-MS, differentiation between Arabica and Robusta was observed for these components, with trigonelline content being higher in Arabica than Robusta [6]. At the same time no visible difference between the nicotinic acid content on regular, decaffeinated and instant coffees was observed.

## Recovery of Bioactive Compounds from Spent Coffee

Spent coffee grounds (SCG), as by-products of coffee beans consumption, contain large amounts of bioactive compounds such as CGAs and derivatives. Increasing awareness for the need of waste reduction and environmental protection has been gaining relevance, as six million tons of residues from the preparation of instant coffee are produced worldwide every year. Also an economically counterpart is not to forget, as the possibility of turning these residues of no commercial value into an economical viable product through a more eco-friendly approach than send it to compost facilities is obviously of interest. Recovery of bioactive compounds into cosmetic, pharmaceutical and food products is a desirable solution at all instances.

The comparison between levels of important hydrophilic antioxidant compounds and the antioxidant capacity on both Robusta and Arabica brews and respective spent coffee grounds has been addressed [167]. 3-, 4-, and 5-monocaffeoylquinic and 3,4-, 3,5-, and 4,5- dicaffeoylquinic acids, caffeine, and browned compounds including melanoidins) and the antioxidant capacity have been addressed using HPLC-DAD and Folin−Ciocalteu, ABTS, DPPH, Fremy's salt, and TEMPO. The antioxidant capacities of the aqueous spent coffee extracts were 46.0−102.3% (filter), <42% (plunger), and 59.2−85.6% (espresso), in comparison to their respective coffee brews. Other report [168] shows that espresso coffee preparation has a reduced extraction efficiency, leaving a significant amount of bioactive compounds retained in grounds after extraction, *e.g.* lipids (12.5%) and CGAs (478.9 mg/100 g), and caffeine (452.6 mg/100 g). Both studies show the great potential and importance of developing ways for expedite analysis and compound recovery.

As the process can only be cost-effective if SCG that contain large amounts of the desired compounds are used, rapid and environmentally friendly analysis techniques are needed for evaluation of the continuous production of residues. NIRS has recently been proposed as a front-end procedure for determination of the antioxidant capacity, the total phenolic and total flavonoid contents of SCG samples [169]. On this study, samples were obtained after the preparation of different volumes of several commercial brands of espresso and dried at 45°C over 1 week. Samples were transferred to flasks and measured in diffuse reflectance mode over a 10000-4000 cm$^{-1}$ range. PLS models were developed for total phenolic and total flavonoid content using reference values from Folin-Ciocaltaeu method, with a coefficient of determination of 0.95. Prediction models developed for antioxidant capacity, using direct and indirect ABTS procedure results as reference values, showed slightly lower coefficient of determination (0.93).

## CONCLUDING REMARKS

Coffee is the most consumed bioactive beverage in the world and the top food product exchanged in the international trade market. Thus, its impact over the world's economy is very important. Its composition varies according to many factors and is responsible for the development of the organoleptic characteristics (aroma, flavor, taste, color) during the roasting process. The compounds showing bioactive properties in coffee beverages are gaining much attention due to their effect on human health and promoting coffee consumption. The most important bioactive compounds are chlorogenic acids, caffeine, trigonelline, melanoidins and diterpenes (kahweol and cafestol). These compounds contribute to the antioxidant capacity of coffee, leading to beneficial effects in several diseases such as T2D, several cancers as well as PD and AD. The need for the study and quantification of bioactive compounds encouraged the development of different methods for their determination. These methods can be categorized into different groups, namely separation-based analytical methods (HPLC and GC coupled with UV-VIS or MS, as well as CE) and non-separative methods such as: electrochemical and spectroscopy techniques. Each method has both advantages and disadvantages, thus the choice depends mostly on the component to be determined and falls ultimately to the investigation team possibilities, availability and purpose.

## CONFLICT OF INTEREST

The authors confirm that they have no conflict of interest to declare for this publication.

## ACKNOWLEDGEMENTS

Mafalda Sarraguça, Miguel Lopo and Ricardo Páscoa acknowledges Fundação para a Ciência a Tecnologia (FCT) for the grant SFRH/ BPD/ 74788/ 2010, SFRH/ BD/ 91521/ 2012 and SFRH/ BPD/ 81384/ 2011, respectively. This work received financial support from the European Union (FEDER funds through COMPETE) and National Funds (FCT, Fundação para a Ciência e Tecnologia) through project UID/QUI/50006/2013.

## REFERENCES

[1]    Murthy PS, Naidu MM. Sustainable management of coffee industry by-products and value addition-A review. Resour Conserv Recycling 2012; 66: 45-58.
[http://dx.doi.org/10.1016/j.resconrec.2012.06.005]

[2]    Butt MS, Sultan MT. Coffee and its consumption: benefits and risks. Crit Rev Food Sci Nutr 2011; 51(4): 363-73.
[http://dx.doi.org/10.1080/10408390903586412] [PMID: 21432699]

[3]     Mussatto SI, Machado EM, Martins S, Teixeira JA. Production, composition, and application of coffee and its industrial residues. Food Bioprocess Tech 2011; 4: 661-72.
[http://dx.doi.org/10.1007/s11947-011-0565-z]

[4]     ICO. International Coffee Organization. Available from: http://www.ico.org 21 April. 2015

[5]     WCT. World Coffee Trade - An Overview The coffee exporter's guide. 3rd., Geneva: International Trade Centre (ITC) 2011.

[6]     Perrone D, Donangelo CM, Farah A. Fast simultaneous analysis of caffeine, trigonelline, nicotinic acid and sucrose in coffee by liquid chromatography-mass spectrometry. Food Chem 2008; 110(4): 1030-5.
[http://dx.doi.org/10.1016/j.foodchem.2008.03.012] [PMID: 26047298]

[7]     Farah A, Monteiro MC, Calado V, Franca AS, Trugo LC. Correlation between cup quality and chemical attributes of Brazilian coffee. Food Chem 2006; 98: 373-80.
[http://dx.doi.org/10.1016/j.foodchem.2005.07.032]

[8]     Charles-Bernard M, Kraehenbuehl K, Rytz A, Roberts DD. Interactions between volatile and nonvolatile coffee components. 1. Screening of nonvolatile components. J Agric Food Chem 2005; 53(11): 4417-25.
[http://dx.doi.org/10.1021/jf048021q] [PMID: 15913304]

[9]     Wagemaker TA, Carvalho CR, Maia NB, Baggio SR, Guerreiro O. Sun protection factor, content and composition of lipid fraction of green coffee beans. Ind Crops Prod 2011; 33: 469-73.
[http://dx.doi.org/10.1016/j.indcrop.2010.10.026]

[10]    Drewnowski A. The science and complexity of bitter taste. Nutr Rev 2001; 59(6): 163-9.
[http://dx.doi.org/10.1111/j.1753-4887.2001.tb07007.x] [PMID: 11444592]

[11]    Higdon JV, Frei B. Coffee and health: a review of recent human research. Crit Rev Food Sci Nutr 2006; 46(2): 101-23.
[http://dx.doi.org/10.1080/10408390500400009] [PMID: 16507475]

[12]    Ky CL, Louarn J, Dussert S, Guyot B, Hamon S, Noirot M. Caffeine, trigonelline, chlorogenic acids and sucrose diversity in wild Coffea arabica L. and C-canephora P. accessions. Food Chem 2001; 75: 223-30.
[http://dx.doi.org/10.1016/S0308-8146(01)00204-7]

[13]    Farah A, Donangelo CM. Phenolic compunds in coffee. Braz J Plant Physiol 2006; 18: 23-6.
[http://dx.doi.org/10.1590/S1677-04202006000100003]

[14]    Clifford MN. Chlorogenic acids and other cinnamates - nature, occurrence and dietary burden. J Sci Food Agric 1999; 79: 362-72.
[http://dx.doi.org/10.1002/(SICI)1097-0010(19990301)79:3<362::AID-JSFA256>3.0.CO;2-D]

[15]    Clifford MN, Knight S, Surucu B, Kuhnert N. Characterization by LC-MS(n) of four new classes of chlorogenic acids in green coffee beans: dimethoxycinnamoylquinic acids, diferuloylquinic acids, caffeoyl-dimethoxycinnamoylquinic acids, and feruloyl-dimethoxycinnamoylquinic acids. J Agric Food Chem 2006; 54(6): 1957-69.
[http://dx.doi.org/10.1021/jf0601665] [PMID: 16536562]

[16]    Monteiro MC, Farah A. Chlorogenic acids in Brazilian Coffea arabica cultivars from various consecutive crops. Food Chem 2012; 134: 611-4.

[http://dx.doi.org/10.1016/j.foodchem.2012.02.118]

[17] Moon JK, Yoo HS, Shibamoto T. Role of roasting conditions in the level of chlorogenic acid content in coffee beans: correlation with coffee acidity. J Agric Food Chem 2009; 57(12): 5365-9.
[http://dx.doi.org/10.1021/jf900012b] [PMID: 19530715]

[18] Ludwig IA, Clifford MN, Lean ME, Ashihara H, Crozier A. Coffee: biochemistry and potential impact on health. Food Funct 2014; 5(8): 1695-717.
[http://dx.doi.org/10.1039/C4FO00042K] [PMID: 24671262]

[19] Feng R, Lu Y, Bowman LL, Qian Y, Castranova V, Ding M. Inhibition of activator protein-1, NF-kappaB, and MAPKs and induction of phase 2 detoxifying enzyme activity by chlorogenic acid. J Biol Chem 2005; 280(30): 27888-95.
[http://dx.doi.org/10.1074/jbc.M503347200] [PMID: 15944151]

[20] Jurkowska RZ, Jurkowski TP, Jeltsch A. Structure and function of mammalian DNA methyltransferases. ChemBioChem 2011; 12(2): 206-22.
[http://dx.doi.org/10.1002/cbic.201000195] [PMID: 21243710]

[21] Park JB. 5-Caffeoylquinic acid and caffeic acid orally administered suppress P-selectin expression on mouse platelets. J Nutr Biochem 2009; 20(10): 800-5.
[http://dx.doi.org/10.1016/j.jnutbio.2008.07.009] [PMID: 18926684]

[22] Matijasevich A, Santos IS, Barros FC. Does caffeine consumption during pregnancy increase the risk of fetal mortality? A literature review. Cad Saude Publica 2005; 21(6): 1676-84.
[http://dx.doi.org/10.1590/S0102-311X2005000600014] [PMID: 16410851]

[23] Mazzafera P, Crozier A, Magalhaes AC. Caffeine metabolism in coffea-arabica and other species of coffee. Phytochemistry 1991; 30: 3913-6.
[http://dx.doi.org/10.1016/0031-9422(91)83433-L]

[24] Belay A, Ture K, Redi M, Asfaw A. Measurement of caffeine in coffee beans with UV/vis spectrometer. Food Chem 2008; 108: 310-5.
[http://dx.doi.org/10.1016/j.foodchem.2007.10.024]

[25] Ferruzzi MG. The influence of beverage composition on delivery of phenolic compounds from coffee and tea. Physiol Behav 2010; 100(1): 33-41.
[http://dx.doi.org/10.1016/j.physbeh.2010.01.035] [PMID: 20138903]

[26] McCusker RR, Goldberger BA, Cone EJ. Caffeine content of specialty coffees. J Anal Toxicol 2003; 27(7): 520-2.
[http://dx.doi.org/10.1093/jat/27.7.520] [PMID: 14607010]

[27] Hečimović I, Belščak-Cvitanović A, Horžić D, Komes D. Comparative study of polyphenols and caffeine in different coffee varieties affected by the degree of roasting. Food Chem 2011; 129(3): 991-1000.
[http://dx.doi.org/10.1016/j.foodchem.2011.05.059] [PMID: 25212328]

[28] Heckman MA, Weil J, Gonzalez de Mejia E. Caffeine (1, 3, 7-trimethylxanthine) in foods: a comprehensive review on consumption, functionality, safety, and regulatory matters. J Food Sci 2010; 75(3): R77-87.
[http://dx.doi.org/10.1111/j.1750-3841.2010.01561.x] [PMID: 20492310]

[29] Borota D, Murray E, Keceli G, *et al.* Post-study caffeine administration enhances memory consolidation in humans. Nat Neurosci 2014; 17(2): 201-3.
[http://dx.doi.org/10.1038/nn.3623] [PMID: 24413697]

[30] Trevitt J, Kawa K, Jalali A, Larsen C. Differential effects of adenosine antagonists in two models of parkinsonian tremor. Pharmacol Biochem Behav 2009; 94(1): 24-9.
[http://dx.doi.org/10.1016/j.pbb.2009.07.001] [PMID: 19602422]

[31] Zhou J, Chan L, Zhou S. Trigonelline: a plant alkaloid with therapeutic potential for diabetes and central nervous system disease. Curr Med Chem 2012; 19(21): 3523-31.
[http://dx.doi.org/10.2174/092986712801323171] [PMID: 22680628]

[32] Nogueira M, Trugo LC. Distribuição de isômeros de ácido clorogênico e teores de cafeína e trigonelina em cafés solúveis brasileiros. Food Sci Technol (Campinas) 2003; 23: 296-9.
[http://dx.doi.org/10.1590/S0101-20612003000200033]

[33] Monteiro MC, Trugo LC. Determination of bioactive compounds in Brazilian roasted coffees. Quim Nova 2005; 28: 637-41.
[http://dx.doi.org/10.1590/S0100-40422005000400016]

[34] Bicho NC, Leitão AE, Ramalho JC, Lidon FC. Identification of chemical clusters discriminators of the roast degree in Arabica and Robusta coffee beans. Eur Food Res Technol 2011; 233: 303-11.
[http://dx.doi.org/10.1007/s00217-011-1518-5]

[35] Zhou J, Zhou S, Zeng S. Experimental diabetes treated with trigonelline: effect on β cell and pancreatic oxidative parameters. Fundam Clin Pharmacol 2013; 27(3): 279-87.
[http://dx.doi.org/10.1111/j.1472-8206.2011.01022.x] [PMID: 22172053]

[36] van Dijk AE, Olthof MR, Meeuse JC, Seebus E, Heine RJ, van Dam RM. Acute effects of decaffeinated coffee and the major coffee components chlorogenic acid and trigonelline on glucose tolerance. Diabetes Care 2009; 32(6): 1023-5.
[http://dx.doi.org/10.2337/dc09-0207] [PMID: 19324944]

[37] Gaur V, Bodhankar SL, Mohan V, Thakurdesai PA. Neurobehavioral assessment of hydroalcoholic extract of Trigonella foenum-graecum seeds in rodent models of Parkinson's disease. Pharm Biol 2013; 51(5): 550-7.
[http://dx.doi.org/10.3109/13880209.2012.747547] [PMID: 23368940]

[38] Hirakawa N, Okauchi R, Miura Y, Yagasaki K. Anti-invasive activity of niacin and trigonelline against cancer cells. Biosci Biotechnol Biochem 2005; 69(3): 653-8.
[http://dx.doi.org/10.1271/bbb.69.653] [PMID: 15785001]

[39] Lamartiniere CA, Moore JB, Brown NM, Thompson R, Hardin MJ, Barnes S. Genistein suppresses mammary cancer in rats. Carcinogenesis 1995; 16(11): 2833-40.
[http://dx.doi.org/10.1093/carcin/16.11.2833] [PMID: 7586206]

[40] Arlt A, Sebens S, Krebs S, *et al.* Inhibition of the Nrf2 transcription factor by the alkaloid trigonelline renders pancreatic cancer cells more susceptible to apoptosis through decreased proteasomal gene expression and proteasome activity. Oncogene 2013; 32(40): 4825-35.
[http://dx.doi.org/10.1038/onc.2012.493] [PMID: 23108405]

[41] Almeida AA, Farah A, Silva DA, Nunan EA, Glória MB. Antibacterial activity of coffee extracts and

selected coffee chemical compounds against enterobacteria. J Agric Food Chem 2006; 54(23): 8738-43.
[http://dx.doi.org/10.1021/jf0617317] [PMID: 17090115]

[42]   Esquivel P, Jimenez VM. Functional properties of coffee and coffee by-products. Food Res Int 2012; 46: 488-95.
[http://dx.doi.org/10.1016/j.foodres.2011.05.028]

[43]   Cosovic B, Vojvodic V, Boskovic N, Plavsic M, Lee C. Characterization of natural and synthetic humic substances (melanoidins) by chemical composition and adsorption measurements. Org Geochem 2010; 41: 200-5.
[http://dx.doi.org/10.1016/j.orggeochem.2009.10.002]

[44]   Kim JS, Lee YS. Enolization and racemization reactions of glucose and fructose on heating with amino-acid enantiomers and the formation of melanoidins as a result of the Maillard reaction. Amino Acids 2009; 36(3): 465-74.
[http://dx.doi.org/10.1007/s00726-008-0104-z] [PMID: 18496645]

[45]   Bekedam EK, Roos E, Schols HA, Van Boekel MA, Smit G. Low molecular weight melanoidins in coffee brew. J Agric Food Chem 2008; 56(11): 4060-7.
[http://dx.doi.org/10.1021/jf8001894] [PMID: 18489118]

[46]   Bekedam EK, De Laat MP, Schols HA, Van Boekel MA, Smit G. Arabinogalactan proteins are incorporated in negatively charged coffee brew melanoidins. J Agric Food Chem 2007; 55(3): 761-8.
[http://dx.doi.org/10.1021/jf063010d] [PMID: 17263472]

[47]   Lindenmeier M, Faist V, Hofmann T. Structural and functional characterization of pronyl-lysine, a novel protein modification in bread crust melanoidins showing *in vitro* antioxidative and phase I/II enzyme modulating activity. J Agric Food Chem 2002; 50(24): 6997-7006.
[http://dx.doi.org/10.1021/jf020618n] [PMID: 12428950]

[48]   Vitaglione P, Fogliano V, Pellegrini N. Coffee, colon function and colorectal cancer. Food Funct 2012; 3(9): 916-22.
[http://dx.doi.org/10.1039/c2fo30037k] [PMID: 22627289]

[49]   Lee K-A, Chae J-I, Shim J-H. Natural diterpenes from coffee, cafestol and kahweol induce apoptosis through regulation of specificity protein 1 expression in human malignant pleural mesothelioma. J Biomed Sci 2012; 19: 60.
[http://dx.doi.org/10.1186/1423-0127-19-60] [PMID: 22734486]

[50]   Acevedo F, Rubilar M, Scheuermann E, Cancino B, Uquiche E, Garces M, *et al.* Spent coffee grounds as a renewable source of bioactive compounds. J Biobased Mater Bioenergy 2013; 7: 420-8.
[http://dx.doi.org/10.1166/jbmb.2013.1369]

[51]   Jeszka-Skowron M, Zgola-Grzeskowiak A, Grzeskowiak T. Analytical methods applied for the characterization and the determination of bioactive compounds in coffee. Eur Food Res Technol 2015; 240: 19-31.
[http://dx.doi.org/10.1007/s00217-014-2356-z]

[52]   deRoos B. vanderWeg G, Urgert R, vandeBovenkamp P, Charrier A, Katan MB. Levels of cafestol, kahweol, and related diterpenoids in wild species of the coffee plant *Coffea*. J Agric Food Chem 1997; 45: 3065-9.

[http://dx.doi.org/10.1021/jf9700900]

[53]    Dias RC, de Faria-Machado AF, Mercadante A, Bragagnolo N, Benassi MD. Roasting process affects the profile of diterpenes in coffee. Eur Food Res Technol 2014; 239: 961-70.
[http://dx.doi.org/10.1007/s00217-014-2293-x]

[54]    Cavin C, Holzhaeuser D, Scharf G, Constable A, Huber WW, Schilter B. Cafestol and kahweol, two coffee specific diterpenes with anticarcinogenic activity. Food Chem Toxicol 2002; 40(8): 1155-63.
[http://dx.doi.org/10.1016/S0278-6915(02)00029-7] [PMID: 12067578]

[55]    Bøhn SK, Blomhoff R, Paur I. Coffee and cancer risk, epidemiological evidence, and molecular mechanisms. Mol Nutr Food Res 2014; 58(5): 915-30.
[http://dx.doi.org/10.1002/mnfr.201300526] [PMID: 24668519]

[56]    George SE, Ramalakshmi K, Mohan Rao LJ. A perception on health benefits of coffee. Crit Rev Food Sci Nutr 2008; 48(5): 464-86.
[http://dx.doi.org/10.1080/10408390701522445] [PMID: 18464035]

[57]    Serafini M, Testa MF. Redox ingredients for oxidative stress prevention: the unexplored potentiality of coffee. Clin Dermatol 2009; 27(2): 225-9.
[http://dx.doi.org/10.1016/j.clindermatol.2008.04.007] [PMID: 19168004]

[58]    Grubben MJ, Van Den Braak CC, Broekhuizen R, *et al.* The effect of unfiltered coffee on potential biomarkers for colonic cancer risk in healthy volunteers: a randomized trial. Aliment Pharmacol Ther 2000; 14(9): 1181-90.
[http://dx.doi.org/10.1046/j.1365-2036.2000.00826.x] [PMID: 10971235]

[59]    Esposito F, Morisco F, Verde V, *et al.* Moderate coffee consumption increases plasma glutathione but not homocysteine in healthy subjects. Aliment Pharmacol Ther 2003; 17(4): 595-601.
[http://dx.doi.org/10.1046/j.1365-2036.2003.01429.x] [PMID: 12622769]

[60]    Natella F, Scaccini C. Role of coffee in modulation of diabetes risk. Nutr Rev 2012; 70(4): 207-17.
[http://dx.doi.org/10.1111/j.1753-4887.2012.00470.x] [PMID: 22458694]

[61]    Zhang Y, Lee ET, Cowan LD, Fabsitz RR, Howard BV. Coffee consumption and the incidence of type 2 diabetes in men and women with normal glucose tolerance: the strong heart study. Nutr Metab Cardiovasc Dis 2011; 21(6): 418-23.
[http://dx.doi.org/10.1016/j.numecd.2009.10.020] [PMID: 20171062]

[62]    van Dam RM, Feskens EJ. Coffee consumption and risk of type 2 diabetes mellitus. Lancet 2002; 360(9344): 1477-8.
[http://dx.doi.org/10.1016/S0140-6736(02)11436-X] [PMID: 12433517]

[63]    Agardh EE, Carlsson S, Ahlbom A, *et al.* Coffee consumption, type 2 diabetes and impaired glucose tolerance in Swedish men and women. J Intern Med 2004; 255(6): 645-52.
[http://dx.doi.org/10.1111/j.1365-2796.2004.01331.x] [PMID: 15147528]

[64]    van Dam RM, Dekker JM, Nijpels G, Stehouwer CD, Bouter LM, Heine RJ. Coffee consumption and incidence of impaired fasting glucose, impaired glucose tolerance, and type 2 diabetes: the Hoorn Study. Diabetologia 2004; 47(12): 2152-9.
[http://dx.doi.org/10.1007/s00125-004-1573-6] [PMID: 15662556]

[65]    Hamer M, Witte DR, Mosdøl A, Marmot MG, Brunner EJ. Prospective study of coffee and tea

consumption in relation to risk of type 2 diabetes mellitus among men and women: the Whitehall II study. Br J Nutr 2008; 100(5): 1046-53.
[http://dx.doi.org/10.1017/S0007114508944135] [PMID: 18315891]

[66]    Bidel S, Silventoinen K, Hu G, Lee DH, Kaprio J, Tuomilehto J. Coffee consumption, serum gamma-glutamyltransferase and risk of type II diabetes. Eur J Clin Nutr 2008; 62(2): 178-85.
[http://dx.doi.org/10.1038/sj.ejcn.1602712] [PMID: 17342160]

[67]    Paynter NP, Yeh HC, Voutilainen S, *et al.* Coffee and sweetened beverage consumption and the risk of type 2 diabetes mellitus: the atherosclerosis risk in communities study. Am J Epidemiol 2006; 164(11): 1075-84.
[http://dx.doi.org/10.1093/aje/kwj323] [PMID: 16982672]

[68]    Pereira MA, Parker ED, Folsom AR. Coffee consumption and risk of type 2 diabetes mellitus: an 11-year prospective study of 28 812 postmenopausal women. Arch Intern Med 2006; 166(12): 1311-6.
[http://dx.doi.org/10.1001/archinte.166.12.1311] [PMID: 16801515]

[69]    Fuhrman BJ, Smit E, Crespo CJ, Garcia-Palmieri MR. Coffee intake and risk of incident diabetes in Puerto Rican men: results from the Puerto Rico Heart Health Program. Public Health Nutr 2009; 12(6): 842-8.
[http://dx.doi.org/10.1017/S1368980008003303] [PMID: 18775084]

[70]    Odegaard AO, Pereira MA, Koh WP, Arakawa K, Lee HP, Yu MC. Coffee, tea, and incident type 2 diabetes: the Singapore Chinese Health Study. Am J Clin Nutr 2008; 88(4): 979-85.
[PMID: 18842784]

[71]    Kato M, Noda M, Inoue M, Kadowaki T, Tsugane S, Grp JS. Psychological factors, coffee and risk of diabetes mellitus among middle-aged Japanese: a population-based prospective study in the JPHC study cohort. Endocr J 2009; 56(3): 459-68.
[http://dx.doi.org/10.1507/endocrj.K09E-003] [PMID: 19270421]

[72]    Oba S, Nagata C, Nakamura K, *et al.* Consumption of coffee, green tea, oolong tea, black tea, chocolate snacks and the caffeine content in relation to risk of diabetes in Japanese men and women. Br J Nutr 2010; 103(3): 453-9.
[http://dx.doi.org/10.1017/S0007114509991966] [PMID: 19818197]

[73]    Rosengren A, Dotevall A, Wilhelmsen L, Thelle D, Johansson S. Coffee and incidence of diabetes in Swedish women: a prospective 18-year follow-up study. J Intern Med 2004; 255(1): 89-95.
[http://dx.doi.org/10.1046/j.1365-2796.2003.01260.x] [PMID: 14687243]

[74]    van Dam RM, Hu FB. Coffee consumption and risk of type 2 diabetes: a systematic review. JAMA 2005; 294(1): 97-104.
[http://dx.doi.org/10.1001/jama.294.1.97] [PMID: 15998896]

[75]    Huxley R, Lee CM, Barzi F, *et al.* Coffee, decaffeinated coffee, and tea consumption in relation to incident type 2 diabetes mellitus: a systematic review with meta-analysis. Arch Intern Med 2009; 169(22): 2053-63.
[http://dx.doi.org/10.1001/archinternmed.2009.439] [PMID: 20008687]

[76]    Greenberg JA, Boozer CN, Geliebter A. Coffee, diabetes, and weight control. Am J Clin Nutr 2006; 84(4): 682-93.
[PMID: 17023692]

[77]　Lenzen S. Oxidative stress: the vulnerable beta-cell. Biochem Soc Trans 2008; 36(Pt 3): 343-7.
　　　[http://dx.doi.org/10.1042/BST0360343] [PMID: 18481954]

[78]　Prasad K, Mantha SV, Muir AD, Westcott ND. Protective effect of secoisolariciresinol diglucoside against streptozotocin-induced diabetes and its mechanism. Mol Cell Biochem 2000; 206(1-2): 141-9.
　　　[http://dx.doi.org/10.1023/A:1007018030524] [PMID: 10839204]

[79]　Kagami K, Morita H, Onda K, Hirano T, Oka K. Protective effect of caffeine on streptozotocin-induced beta-cell damage in rats. J Pharm Pharmacol 2008; 60(9): 1161-5.
　　　[http://dx.doi.org/10.1211/jpp.60.9.0007] [PMID: 18718119]

[80]　Fund WC. Food, Nutrition, Physical activity and the prevention of cancer: a global perspective. Washington, DC: AICR 2007.

[81]　Li XJ, Ren ZJ, Qin JW, *et al.* Coffee consumption and risk of breast cancer: an up-to-date meta-analysis. PLoS One 2013; 8(1): e52681.
　　　[http://dx.doi.org/10.1371/journal.pone.0052681] [PMID: 23308117]

[82]　Jiang W, Wu Y, Jiang X. Coffee and caffeine intake and breast cancer risk: an updated dose-response meta-analysis of 37 published studies. Gynecol Oncol 2013; 129(3): 620-9.
　　　[http://dx.doi.org/10.1016/j.ygyno.2013.03.014] [PMID: 23535278]

[83]　Yu X, Bao Z, Zou J, Dong J. Coffee consumption and risk of cancers: a meta-analysis of cohort studies. BMC Cancer 2011; 11: 96.
　　　[http://dx.doi.org/10.1186/1471-2407-11-96] [PMID: 21406107]

[84]　Tang N, Zhou B, Wang B, Yu R. Coffee consumption and risk of breast cancer: a metaanalysis. Am J Obstet Gynecol 2009; 200(3): 290.e1-9.
　　　[http://dx.doi.org/10.1016/j.ajog.2008.10.019] [PMID: 19114275]

[85]　Lowcock EC, Cotterchio M, Anderson LN, Boucher BA, El-Sohemy A. High coffee intake, but not caffeine, is associated with reduced estrogen receptor negative and postmenopausal breast cancer risk with no effect modification by CYP1A2 genotype. Nutr Cancer 2013; 65(3): 398-409.
　　　[http://dx.doi.org/10.1080/01635581.2013.768348] [PMID: 23530639]

[86]　Simonsson M, Söderlind V, Henningson M, *et al.* Coffee prevents early events in tamoxifen-treated breast cancer patients and modulates hormone receptor status. Cancer Causes Control 2013; 24(5): 929-40.
　　　[http://dx.doi.org/10.1007/s10552-013-0169-1] [PMID: 23412805]

[87]　Bravi F, Scotti L, Bosetti C, *et al.* Food groups and endometrial cancer risk: a case-control study from Italy. Am J Obstet Gynecol 2009; 200(3): 293.e1-7.
　　　[http://dx.doi.org/10.1016/j.ajog.2008.09.015] [PMID: 19091304]

[88]　Koizumi T, Nakaya N, Okamura C, *et al.* Case-control study of coffee consumption and the risk of endometrial endometrioid adenocarcinoma. Eur J Cancer Prev 2008; 17(4): 358-63.
　　　[http://dx.doi.org/10.1097/CEJ.0b013e3282f0c02c] [PMID: 18562962]

[89]　Hirose K, Niwa Y, Wakai K, Matsuo K, Nakanishi T, Tajima K. Coffee consumption and the risk of endometrial cancer: Evidence from a case-control study of female hormone-related cancers in Japan. Cancer Sci 2007; 98(3): 411-5.
　　　[http://dx.doi.org/10.1111/j.1349-7006.2007.00391.x] [PMID: 17270030]

[90] Shimazu T, Inoue M, Sasazuki S, *et al.* Coffee consumption and risk of endometrial cancer: a prospective study in Japan. Int J Cancer 2008; 123(10): 2406-10.
[http://dx.doi.org/10.1002/ijc.23760] [PMID: 18711700]

[91] Wilson KM, Kasperzyk JL, Rider JR, *et al.* Coffee consumption and prostate cancer risk and progression in the Health Professionals Follow-up Study. J Natl Cancer Inst 2011; 103(11): 876-84.
[http://dx.doi.org/10.1093/jnci/djr151] [PMID: 21586702]

[92] Wilson KM, Bälter K, Möller E, *et al.* Coffee and risk of prostate cancer incidence and mortality in the Cancer of the Prostate in Sweden Study. Cancer Causes Control 2013; 24(8): 1575-81.
[http://dx.doi.org/10.1007/s10552-013-0234-9] [PMID: 23702886]

[93] Geybels MS, Neuhouser ML, Wright JL, Stott-Miller M, Stanford JL. Coffee and tea consumption in relation to prostate cancer prognosis. Cancer Causes Control 2013; 24(11): 1947-54.
[http://dx.doi.org/10.1007/s10552-013-0270-5] [PMID: 23907772]

[94] Sang LX, Chang B, Li XH, Jiang M. Consumption of coffee associated with reduced risk of liver cancer: a meta-analysis. BMC Gastroenterol 2013; 13: 34.
[http://dx.doi.org/10.1186/1471-230X-13-34] [PMID: 23433483]

[95] Larsson SC, Wolk A. Coffee consumption and risk of liver cancer: a meta-analysis. Gastroenterology 2007; 132(5): 1740-5.
[http://dx.doi.org/10.1053/j.gastro.2007.03.044] [PMID: 17484871]

[96] Bravi F, Bosetti C, Tavani A, *et al.* Coffee drinking and hepatocellular carcinoma risk: a meta-analysis. Hepatology 2007; 46(2): 430-5.
[http://dx.doi.org/10.1002/hep.21708] [PMID: 17580359]

[97] Lai GY, Weinstein SJ, Albanes D, *et al.* The association of coffee intake with liver cancer incidence and chronic liver disease mortality in male smokers. Br J Cancer 2013; 109(5): 1344-51.
[http://dx.doi.org/10.1038/bjc.2013.405] [PMID: 23880821]

[98] Tavani A, La Vecchia C. Coffee, decaffeinated coffee, tea and cancer of the colon and rectum: a review of epidemiological studies, 1990-2003. Cancer Causes Control 2004; 15(8): 743-57.
[http://dx.doi.org/10.1023/B:CACO.0000043415.28319.c1] [PMID: 15456988]

[99] Li G, Ma D, Zhang Y, Zheng W, Wang P. Coffee consumption and risk of colorectal cancer: a meta-analysis of observational studies. Public Health Nutr 2013; 16(2): 346-57.
[http://dx.doi.org/10.1017/S1368980012002601] [PMID: 22694939]

[100] Galeone C, Turati F, La Vecchia C, Tavani A. Coffee consumption and risk of colorectal cancer: a meta-analysis of case-control studies. Cancer Causes Control 2010; 21(11): 1949-59.
[http://dx.doi.org/10.1007/s10552-010-9623-5] [PMID: 20680435]

[101] Tian C, Wang W, Hong Z, Zhang X. Coffee consumption and risk of colorectal cancer: a dose-response analysis of observational studies. Cancer Causes Control 2013; 24(6): 1265-8.
[http://dx.doi.org/10.1007/s10552-013-0200-6] [PMID: 23546611]

[102] Sugiyama K, Kuriyama S, Akhter M, *et al.* Coffee consumption and mortality due to all causes, cardiovascular disease, and cancer in Japanese women. J Nutr 2010; 140(5): 1007-13.
[http://dx.doi.org/10.3945/jn.109.109314] [PMID: 20335629]

[103]  Wang ZJ, Ohnaka K, Morita M, *et al.* Dietary polyphenols and colorectal cancer risk: the Fukuoka colorectal cancer study. World J Gastroenterol 2013; 19(17): 2683-90.
[http://dx.doi.org/10.3748/wjg.v19.i17.2683] [PMID: 23674876]

[104]  Mathers JC, Strathdee G, Relton CL. Induction of epigenetic alterations by dietary and other environmental factors. In: Herceg Z, Ushijima T, Eds. Epigenetics and Cancer, Pt B2010. 3-39.
[http://dx.doi.org/10.1016/B978-0-12-380864-6.00001-8]

[105]  Kempf K, Herder C, Erlund I, *et al.* Effects of coffee consumption on subclinical inflammation and other risk factors for type 2 diabetes: a clinical trial. Am J Clin Nutr 2010; 91(4): 950-7.
[http://dx.doi.org/10.3945/ajcn.2009.28548] [PMID: 20181814]

[106]  Mursu J, Voutilainen S, Nurmi T, *et al.* The effects of coffee consumption on lipid peroxidation and plasma total homocysteine concentrations: a clinical trial. Free Radic Biol Med 2005; 38(4): 527-34.
[http://dx.doi.org/10.1016/j.freeradbiomed.2004.11.025] [PMID: 15649655]

[107]  Mantovani A, Allavena P, Sica A, Balkwill F. Cancer-related inflammation. Nature 2008; 454(7203): 436-44.
[http://dx.doi.org/10.1038/nature07205] [PMID: 18650914]

[108]  Karin M, Greten FR. NF-kappaB: linking inflammation and immunity to cancer development and progression. Nat Rev Immunol 2005; 5(10): 749-59.
[http://dx.doi.org/10.1038/nri1703] [PMID: 16175180]

[109]  Conney AH, Lou YR, Nghiem P, Bernard JJ, Wagner GC, Lu YP. Inhibition of UVB-induced nonmelanoma skin cancer: a path from tea to caffeine to exercise to decreased tissue fat. Top Curr Chem 2013; 329: 61-72.
[http://dx.doi.org/10.1007/128_2012_336] [PMID: 22752580]

[110]  Ong KW, Hsu A, Tan BK. Chlorogenic acid stimulates glucose transport in skeletal muscle *via* AMPK activation: a contributor to the beneficial effects of coffee on diabetes. PLoS One 2012; 7(3): e32718.
[http://dx.doi.org/10.1371/journal.pone.0032718] [PMID: 22412912]

[111]  Hwang YP, Jeong HG. The coffee diterpene kahweol induces heme oxygenase-1 *via* the PI3K and p38/Nrf2 pathway to protect human dopaminergic neurons from 6-hydroxydopamine-derived oxidative stress. FEBS Lett 2008; 582(17): 2655-62.
[http://dx.doi.org/10.1016/j.febslet.2008.06.045] [PMID: 18593583]

[112]  Loftfield E, Freedman ND, Graubard BI, *et al.* Coffee drinking and cutaneous melanoma risk in the NIH-AARP diet and health study. J Natl Cancer Inst 2015; 107(2): dju421.
[http://dx.doi.org/10.1093/jnci/dju421] [PMID: 25604135]

[113]  Hernán MA, Takkouche B, Caamaño-Isorna F, Gestal-Otero JJ. A meta-analysis of coffee drinking, cigarette smoking, and the risk of Parkinson's disease. Ann Neurol 2002; 52(3): 276-84.
[http://dx.doi.org/10.1002/ana.10277] [PMID: 12205639]

[114]  Ascherio A, Chen H, Schwarzschild MA, Zhang SM, Colditz GA, Speizer FE. Caffeine, postmenopausal estrogen, and risk of Parkinson's disease. Neurology 2003; 60(5): 790-5.
[http://dx.doi.org/10.1212/01.WNL.0000046523.05125.87] [PMID: 12629235]

[115]  Ascherio A, Weisskopf MG, O'Reilly EJ, *et al.* Coffee consumption, gender, and Parkinson's disease mortality in the cancer prevention study II cohort: the modifying effects of estrogen. Am J Epidemiol

2004; 160(10): 977-84.
[http://dx.doi.org/10.1093/aje/kwh312] [PMID: 15522854]

[116] Chen JF, Xu K, Petzer JP, *et al.* Neuroprotection by caffeine and A(2A) adenosine receptor inactivation in a model of Parkinson's disease. J Neurosci 2001; 21(10): RC143.
[PMID: 11319241]

[117] Ross GW, Abbott RD, Petrovitch H, *et al.* Association of coffee and caffeine intake with the risk of Parkinson disease. JAMA 2000; 283(20): 2674-9.
[http://dx.doi.org/10.1001/jama.283.20.2674] [PMID: 10819950]

[118] Carman AJ, Dacks PA, Lane RF, Shineman DW, Fillit HM. Current evidence for the use of coffee and caffeine to prevent age-related cognitive decline and Alzheimer's disease. J Nutr Health Aging 2014; 18(4): 383-92.
[http://dx.doi.org/10.1007/s12603-014-0021-7] [PMID: 24676319]

[119] Ritchie K, Artero S, Portet F, *et al.* Caffeine, cognitive functioning, and white matter lesions in the elderly: establishing causality from epidemiological evidence. J Alzheimers Dis 2010; 20 (Suppl. 1): S161-6.
[PMID: 20164564]

[120] Ritchie K, Carrière I, de Mendonca A, *et al.* The neuroprotective effects of caffeine: a prospective population study (the Three City Study). Neurology 2007; 69(6): 536-45.
[http://dx.doi.org/10.1212/01.wnl.0000266670.35219.0c] [PMID: 17679672]

[121] Santos C, Lunet N, Azevedo A, de Mendonça A, Ritchie K, Barros H. Caffeine intake is associated with a lower risk of cognitive decline: a cohort study from Portugal. J Alzheimers Dis 2010; 20 (Suppl. 1): S175-85.
[PMID: 20182036]

[122] van Gelder BM, Buijsse B, Tijhuis M, *et al.* Coffee consumption is inversely associated with cognitive decline in elderly European men: the FINE Study. Eur J Clin Nutr 2007; 61(2): 226-32.
[http://dx.doi.org/10.1038/sj.ejcn.1602495] [PMID: 16929246]

[123] Vercambre MN, Berr C, Ritchie K, Kang JH. Caffeine and cognitive decline in elderly women at high vascular risk. J Alzheimers Dis 2013; 35(2): 413-21.
[PMID: 23422357]

[124] Johnson-Kozlow M, Kritz-Silverstein D, Barrett-Connor E, Morton D. Coffee consumption and cognitive function among older adults. Am J Epidemiol 2002; 156(9): 842-50.
[http://dx.doi.org/10.1093/aje/kwf119] [PMID: 12397002]

[125] Eskelinen MH, Ngandu T, Tuomilehto J, Soininen H, Kivipelto M. Midlife coffee and tea drinking and the risk of late-life dementia: a population-based CAIDE study. J Alzheimers Dis 2009; 16(1): 85-91.
[PMID: 19158424]

[126] Lindsay J, Laurin D, Verreault R, *et al.* Risk factors for Alzheimer's disease: a prospective analysis from the Canadian Study of Health and Aging. Am J Epidemiol 2002; 156(5): 445-53.
[http://dx.doi.org/10.1093/aje/kwf074] [PMID: 12196314]

[127] Maia L, de Mendonça A. Does caffeine intake protect from Alzheimer's disease? Eur J Neurol 2002; 9(4): 377-82.

[http://dx.doi.org/10.1046/j.1468-1331.2002.00421.x] [PMID: 12099922]

[128]　Santos C, Costa J, Santos J, Vaz-Carneiro A, Lunet N. Caffeine intake and dementia: systematic review and meta-analysis. J Alzheimers Dis 2010; 20 (Suppl. 1): S187-204.
[PMID: 20182026]

[129]　Nuhu AA. Bioactive micronutrients in coffee: recent analytical approaches for characterization and quantification. ISRN Nutr 2014; 2014: 384230.
[http://dx.doi.org/10.1155/2014/384230] [PMID: 24967266]

[130]　Bizzotto CS, Meinhart AD, Ballus CA, Ghiselli G, Godoy HT. Comparison of capillary electrophoresis and high performance liquid chromatography methods for caffeine determination in decaffeinated coffee. Food Sci Tech-Brazil 2013; 33: 186-91.

[131]　Ziyatdinova G, Nizamova A, Budnikov H. Novel coulometric approach to evaluation of total free polyphenols in tea and coffee beverages in presence of milk proteins. Food Anal Method 2011; 4: 334-40.
[http://dx.doi.org/10.1007/s12161-010-9174-0]

[132]　Yardım Y. Electrochemical behavior of chlorogenic acid at a boron-doped diamond electrode and estimation of the antioxidant capacity in the coffee samples based on its oxidation peak. J Food Sci 2012; 77(4): C408-13.
[http://dx.doi.org/10.1111/j.1750-3841.2011.02609.x] [PMID: 22394181]

[133]　Mersal GA. Experimental and computational studies on the electrochemical oxidation of caffeine at pseudo carbon paste electrode and its voltammetric determination in different real samples. Food Anal Method 2012; 5: 520-9.
[http://dx.doi.org/10.1007/s12161-011-9269-2]

[134]　Santos WdeJ, Santhiago M, Yoshida IV, Kubota LT. Novel electrochemical sensor for the selective recognition of chlorogenic acid. Anal Chim Acta 2011; 695(1-2): 44-50.
[http://dx.doi.org/10.1016/j.aca.2011.03.018] [PMID: 21601028]

[135]　Campa C, Doulbeau S, Dussert S, Hamon S, Noirot M. Qualitative relationship between caffeine nd chlorogenic acid contents among wild Coffea species. Food Chem 2005; 93: 135-9.
[http://dx.doi.org/10.1016/j.foodchem.2004.10.015]

[136]　Trugo LC, Macrae R. A Study of the effect of roasting on the chlorogenic acid composition of coffee using HPLC. Food Chem 1984; 15: 219-27.
[http://dx.doi.org/10.1016/0308-8146(84)90006-2]

[137]　Rodrigues NP, Bragagnolo N. Identification and quantification of bioactive compounds in coffee brews by HPLC-DAD-MSn. J Food Compos Anal 2013; 32: 105-15.
[http://dx.doi.org/10.1016/j.jfca.2013.09.002]

[138]　Ky CL, Noirot M, Hamon S. Comparison of five purification methods for chlorogenic acids in green coffee beans (Coffea sp.). J Agric Food Chem 1997; 45: 786-90.
[http://dx.doi.org/10.1021/jf9605254]

[139]　Budryn G, Nebesny E, Podsedek A, Zyzelewicz D, Materska M, Jankowski S, *et al.* Effect of different extraction methods on the recovery of chlorogenic acids, caffeine and Maillard reaction products in coffee beans. Eur Food Res Technol 2009; 228: 913-22.

[http://dx.doi.org/10.1007/s00217-008-1004-x]

[140] Upadhyay R, Ramalakshmi K, Rao LJ. Microwave-assisted extraction of chlorogenic acids from green coffee beans. Food Chem 2012; 130: 184-8.
[http://dx.doi.org/10.1016/j.foodchem.2011.06.057]

[141] Clifford MN, Johnston KL, Knight S, Kuhnert N. Hierarchical scheme for LC-MSn identification of chlorogenic acids. J Agric Food Chem 2003; 51(10): 2900-11.
[http://dx.doi.org/10.1021/jf026187q] [PMID: 12720369]

[142] Farah A, de Paulis T, Trugo LC, Martin PR. Effect of roasting on the formation of chlorogenic acid lactones in coffee. J Agric Food Chem 2005; 53(5): 1505-13.
[http://dx.doi.org/10.1021/jf048701t] [PMID: 15740032]

[143] Vignoli JA, Viegas MC, Bassoli DG. Benassi MdT. Roasting process affects differently the bioactive compounds and the antioxidant activity of arabica and robusta coffees. Food Res Int 2014; 61: 279-85.
[http://dx.doi.org/10.1016/j.foodres.2013.06.006]

[144] Perrone D, Farah A, Donangelo CM, de Paulis T, Martin PR. Comprehensive analysis of major and minor chlorogenic acids and lactones in economically relevant Brazilian coffee cultivars. Food Chem 2008; 106: 859-67.
[http://dx.doi.org/10.1016/j.foodchem.2007.06.053]

[145] Niseteo T, Komes D, Belščak-Cvitanović A, Horžić D, Budeč M. Bioactive composition and antioxidant potential of different commonly consumed coffee brews affected by their preparation technique and milk addition. Food Chem 2012; 134(4): 1870-7.
[http://dx.doi.org/10.1016/j.foodchem.2012.03.095] [PMID: 23442632]

[146] Fernandes SC, Moccelini SK, Scheeren CW, *et al.* Biosensor for chlorogenic acid based on an ionic liquid containing iridium nanoparticles and polyphenol oxidase. Talanta 2009; 79(2): 222-8.
[http://dx.doi.org/10.1016/j.talanta.2009.03.039] [PMID: 19559869]

[147] Caprioli G, Cortese M, Maggi F, *et al.* Quantification of caffeine, trigonelline and nicotinic acid in espresso coffee: the influence of espresso machines and coffee cultivars. Int J Food Sci Nutr 2014; 65(4): 465-9.
[http://dx.doi.org/10.3109/09637486.2013.873890] [PMID: 24467514]

[148] Fujioka K, Shibamoto T. Chlorogenic acid and caffeine contents in various commercial brewed coffees. Food Chem 2008; 106: 217-21.
[http://dx.doi.org/10.1016/j.foodchem.2007.05.091]

[149] Rodrigues CI, Marta L, Maia R, Miranda M, Ribeirinho M, Maguas C. Application of solid-phase extraction to brewed coffee caffeine and organic acid determination by UV/HPLC. J Food Compos Anal 2007; 20: 440-8.
[http://dx.doi.org/10.1016/j.jfca.2006.08.005]

[150] Moreno FL, Raventos M, Hernandez E, Ruiz Y. Block freeze-concentration of coffee extract: Effect of freezing and thawing stages on solute recovery and bioactive compounds. J Food Eng 2014; 120: 158-66.
[http://dx.doi.org/10.1016/j.jfoodeng.2013.07.034]

[151] Nicolau de Souza RM, Baptista Canuto GA, Eloy Dias RC. Benassi MdT. Levels of bioactive

compounds in commercial roasted and ground coffees. Quim Nova 2010; 33: 885-90.

[152]	Singh BR, Wechter MA, Hu YH, Lafontaine C. Determination of caffeine content in coffee using Fourier transform infra-red spectroscopy in combination with attenuated total reflectance technique: a bioanalytical chemistry experiment for biochemists. Biochem Educ 1998; 26: 243-7.
[http://dx.doi.org/10.1016/S0307-4412(98)00078-8]

[153]	Zhang X, Li W, Yin B, *et al.* Improvement of near infrared spectroscopic (NIRS) analysis of caffeine in roasted Arabica coffee by variable selection method of stability competitive adaptive reweighted sampling (SCARS). Spectrochim Acta A Mol Biomol Spectrosc 2013; 114: 350-6.
[http://dx.doi.org/10.1016/j.saa.2013.05.053] [PMID: 23786975]

[154]	Downey G, Boussion J. Authentication of coffee bean variety by near-infrared reflectance spectroscopy of dried extract. J Sci Food Agric 1996; 71: 41-9.
[http://dx.doi.org/10.1002/(SICI)1097-0010(199605)71:1<41::AID-JSFA546>3.0.CO;2-I]

[155]	Huck CW, Guggenbichler W, Bonn GK. Analysis of caffeine, theobromine and theophylline in coffee by near infrared spectroscopy (NIRS) compared to high-performance liquid chromatography (HPLC) coupled to mass spectrometry. Anal Chim Acta 2005; 538: 195-203.
[http://dx.doi.org/10.1016/j.aca.2005.01.064]

[156]	Ribeiro JS, Ferreira MM, Salva TJ. Chemometric models for the quantitative descriptive sensory analysis of Arabica coffee beverages using near infrared spectroscopy. Talanta 2011; 83(5): 1352-8.
[http://dx.doi.org/10.1016/j.talanta.2010.11.001] [PMID: 21238720]

[157]	Yardım Y, Keskin E, Şentürk Z. Voltammetric determination of mixtures of caffeine and chlorogenic acid in beverage samples using a boron-doped diamond electrode. Talanta 2013; 116: 1010-7.
[http://dx.doi.org/10.1016/j.talanta.2013.08.005] [PMID: 24148509]

[158]	Pacetti D, Boselli E, Balzano M, Frega NG. Authentication of Italian Espresso coffee blends through the GC peak ratio between kahweol and 16-O-methylcafestol. Food Chem 2012; 135(3): 1569-74.
[http://dx.doi.org/10.1016/j.foodchem.2012.06.007] [PMID: 22953895]

[159]	Chartier A, Beaumesnil M, de Oliveira AL, Elfakir C, Bostyn S. Optimization of the isolation and quantitation of kahweol and cafestol in green coffee oil. Talanta 2013; 117: 102-11.
[http://dx.doi.org/10.1016/j.talanta.2013.07.053] [PMID: 24209317]

[160]	Urgert R, Vanderweg G, Kosmeijerschuil TG, Vandebovenkamp P, Hovenier R, Katan MB. Levels of the cholesterol-elevating diterpenes cafestol and kahweol in various coffee brews. J Agric Food Chem 1995; 43: 2167-72.
[http://dx.doi.org/10.1021/jf00056a039]

[161]	Gross G, Jaccaud E, Huggett AC. Analysis of the content of the diterpenes cafestol and kahweol in coffee brews. Food Chem Toxicol 1997; 35(6): 547-54.
[http://dx.doi.org/10.1016/S0278-6915(96)00123-8] [PMID: 9225012]

[162]	Kolling-Speer I, Strohschneider S, Speer K. Determination of free diterpenes in green and roasted coffees. Hrc-J High Res Chrom 1999; 22: 43-6.
[http://dx.doi.org/10.1002/(SICI)1521-4168(19990101)22:1<43::AID-JHRC43>3.0.CO;2-P]

[163]	Araujo JM, Sandi D. Extraction of coffee diterpenes and coffee oil using supercritical carbon dioxide. Food Chem 2007; 101: 1087-94.

[http://dx.doi.org/10.1016/j.foodchem.2006.03.008]

[164] Scharnhop H, Winterhalter P. Isolation of coffee diterpenes by means of high-speed countercurrent chromatography. J Food Compos Anal 2009; 22: 233-7.
[http://dx.doi.org/10.1016/j.jfca.2008.10.018]

[165] Toci A, Farah A, Trugo LC. Effect of decaffeination using dichloromethane on the chemical composition of arabica and robusta raw and roasted coffees. Quim Nova 2006; 29: 965-71.
[http://dx.doi.org/10.1590/S0100-40422006000500015]

[166] DeMaria CA, Trugo LC, Neto FR, Moreira RF, Alviano CS. Composition of green coffee water-soluble fractions and identification of volatiles formed during roasting. Food Chem 1996; 55: 203-7.
[http://dx.doi.org/10.1016/0308-8146(95)00104-2]

[167] Bravo J, Juániz I, Monente C, *et al.* Evaluation of spent coffee obtained from the most common coffeemakers as a source of hydrophilic bioactive compounds. J Agric Food Chem 2012; 60(51): 12565-73.
[http://dx.doi.org/10.1021/jf3040594] [PMID: 23214450]

[168] Cruz R, Cardoso MM, Fernandes L, *et al.* Espresso coffee residues: a valuable source of unextracted compounds. J Agric Food Chem 2012; 60(32): 7777-84.
[http://dx.doi.org/10.1021/jf3018854] [PMID: 22812683]

[169] Pascoa RN, Magalhaes LM, Lopes JA. FT-NIR spectroscopy as a tool for valorization of spent coffee grounds: Application to assessment of antioxidant properties. Food Res Int 2013; 51: 579-86.
[http://dx.doi.org/10.1016/j.foodres.2013.01.035]

**CHAPTER 9**

# Bioactive Compounds of Rice as Health Promoters

## Charu Lata Mahanta[*], Sangeeta Saikia

*Department of Food Engineering and Technology, Tezpur University, Assam, India*

**Abstract:** Rice (*Oryza sativa* L.) is one of the major cereal food crops in the world. It serves as a principal source of carbohydrate. Rice is also traditionally used for therapeutic purposes. The health promoting properties of brown rice has generated research interest. The health promoting properties are attributed to the bioactive compounds in the rice bran. These include vitamins, minerals, phenolic compounds, γ - oryzanol, tocopherols, tocotrienols, anthocyanins, phytosterols and dietary fibre. The concentration of bioactive properties is more in pigmented rice varieties. The bioactives in rice impart cholesterol reduction, anti-inflammatory, anti-tumor, anti-diabetic and anti-oxidative effects. Ferulic acid is the major phenolic acid in rice whose esters predominantly form γ-oryzanol and remains in the human blood stream for longer time than other phenolics thereby offering greater radical scavenging efficiency. The rice bran anthocyanin pigments contribute to the total phenolic and antioxidant properties of rice to a notable extent. The tocols also are potent antioxidants. Both in vivo and in vitro experiments have ascertained the effect of these compounds in animals and human beings. The bioactive potential of rice has been found to widely vary with variety, extent of polishing and processing. Cooking steaming, parboiling, fermentation and germination of rice result in marked changes in proportion and activities of different bioactive components. This chapter focuses on the bioactive compounds in rice, the potential health benefits of rice consumption, and processing effects on bioactivity.

**Keywords:** Anti-cancer, Anthocyanin, Anti-inflammatory, Antioxidant, Bioactive compounds, Bran, Cooking, Fermentation, Ferulic acid, Non-pigmented rice, γ oryzanol, Parboiling, Phenolics, Pigmented rice, Plasma cholesterol, Rice, Therapeutic, Tocopherols, Tocotrienols, Tricin.

[*] **Corresponding author Charu Lata Mahanta:** Department of Food Engineering and Technology, School of Engineering, Napaam-784028, Tezpur University, Assam, India; Tel: +91-3712-275702; Fax: +91-3712-267006; Email: charu@tezu.ernet.in.

## INTRODUCTION

Rice *(Oryza sativa)* is a leading cereal crop and staple food to more than half of the world's population. The harvested paddy undergoes milling and polishing to yield white rice. The first step of milling consists of hulling *i.e.* removal of the outer layer or hull to yield brown rice. The brown rice is further milled to remove the bran layer and germ. The resulting white rice is then polished and graded. The appearance and flavour of rice improves on polishing. However, compared to the brown rice, it has lower nutritional and therapeutic value as the bran layer that is removed on polishing is rich in vitamins, phenolic compounds, minerals, dietary fibre and other bioactive compounds such as γ-oryzanol, tocopherols, tocotrienols, and phytosterols [1, 2]. Bioactive compounds are the non-nutrient secondary metabolites having health promoting functional properties.

Rice varieties are classified into two broad categories as non-pigmented and pigmented rice depending on their bran colour. Non-pigmented rices have brown coloured bran and are commonly grown worldwide. Varieties with black, red and purple coloured bran are classified as pigmented rice varieties. Until recently, the pigmented rice varieties were being cultivated in a very limited quantity in isolated places of the world either for making specific traditional foods or beverage products. Nowadays, the pigmented rices are gaining popularity owing to their many therapeutic uses [3] and growing consumer awareness about the goodness of the rice. The content of bioactive compounds is more in pigmented rices [4, 5] and therapeutic properties are attributed to these compounds.

### Traditional Therapeutic Uses of Rice

The traditional medicinal system in many Asian countries recognises the therapeutic properties of rice [6]. From ancient times, rice is used in the treatment of number of gastrointestinal disorders, diabetes, high cholesterol conditions, skin disorders, jaundice, *etc.* [6 - 8]. In traditional Chinese medicine, pigmented rice has been used for strengthening kidney function, treating anaemia, promoting blood circulation, removing blood stasis, improving blood flow, treating diabetes, and strengthening sight [9]. Red yeast rice, a fermented rice product, is used in traditional Chinese medicine for its cholesterol lowering properties [10]. In China, dried and sprouted rice is used as an aid for digestion. Boiled rice is used to treat acute inflammation of the inner body tissues in many Asian countries. Similarly, rice bran extract is used to prevent and cure beriberi, a disease caused due to Vitamin B-1 (thiamine) deficiency. A variety of black rice, Njavara, cultivated in Kerala, India is referred to in Ayurveda for the treatment of rheumatoid arthritis, paralysis, neurodegenerative diseases and in rejuvenation therapy [11]. Rakthashali, a red rice variety of India, besides its nutritional significance is used in various medications like in allergies, skin ailments, uterus related problems, nerve disorders, gastro-intestinal problems, liver, kidney disorders, fever, infections and in promoting lactation [12].

## Composition and Antioxidant Effects of the Bioactive Compounds in Rice

The bioactive compounds are concentrated in the outer bran layers of the rice grain [2, 6]. Bioactive compounds in rice include phenolic compounds, $\gamma$-oryzanol, tocopherols, tocotrienols, $\beta$-sitosterol, $\gamma$- campesterol and vitamins like thiamine, riboflavin and niacin. Bran from thicker kernels (>1.84 mm) contains a higher content of oryzanol, tocopherols, and tocotrienols [13] that have strong antioxidant properties. $\gamma$-tocotrienol contributes the most to the total tocol content (27–63%), followed by $\alpha$-tocopherol (10–30%), $\alpha$-tocotrienol (9–19%), $\gamma$-tocopherol (9–14%), $\delta$-tocotrienol (2–6%), $\beta$-tocotrienol (1–4%), $\beta$-tocopherol (1–2%), and $\delta$-tocopherol (1–2%) [3]. Japonica varieties contain a higher content of total tocopherol, total tocotrienol and $\gamma$-oryzanol than indica varieties [14]. The phenolic compounds in rice comprise of phenolic acids and flavonoids. The dominant phenolic acids in rice are ferulic acid and *p*-coumaric acid [15]. Other phenolic acids include sinapic, protocatechuic, chlorogenic, vanillic, syringic, caffeic and gallic acids [16]. Among the four types of rice ranked by color, black rice varieties exhibit the highest antioxidant activities, followed by purple, red, and brown rice varieties. Furthermore, insoluble compounds appear to constitute the major fraction of phenolic acids and proanthocyanidins in rice, but not of flavonoids and anthocyanins [2, 3]. Njavara, the medicinal rice variety contains about 50% higher soluble and insoluble forms of ferulic ester than non-medicinal rice [2]. Among the flavonoids, tricin, tricinin and anthocyanins are prominent. The pigmented varieties contain higher levels of anthocyanins than the non pigmented rice. Anthocyanins are group of red to purple water soluble flavonoids.

The red rice is rich in proanthocyanidins, while the black rice anthocyanins predominantly includes cyanidin-3-*O*-*β*-D-glucoside, cyanidin-3, 5-diglucoside, peonidin-3-*O*-glucoside, and pelargonidin-3, 5-diglucoside [17]. Trans-β-carotene, quercetin and isorhamnetin were reported to be present within the range of 33.60–41.0, 1.08–2.85 and 0.05–0.83 μg/g, respectively in the bran of Thai black rice cultivars [18].

Black-purple rice is becoming popular with health conscious food consumers. HPLC–PDA–MS study of dehulled black–purple rice variety Asamurasaki [19] revealed a high concentration of seven anthocyanins (1400 μg/g fresh weight) with cyanidin-3-*O*-glucoside and peonidin-3-*O*-glucoside predominating. Five flavonol glycosides, principally quercetin-3-*O*-glucoside and quercetin-3-*O*-rutinoside, and flavones were detected at a total concentration of 189 μg/g. The grains also contained 3.9 μg/g of carotenoids comprising of lutein, zeaxanthin, lycopene and β-carotene. γ-Oryzanol (279 μg/g) was also present as a mixture of 24-methylenecycloartenol ferulate, campesterol ferulate, cycloartenol ferulate and β-sitosterol ferulate [19]. These ferulate esters are the major oryzanols in rice bran [20]. Nystrom *et al.* [21] suggested that 4-hydroxyl moiety of *trans*-ferulic acid might be the active centre of γ-oryzanol activity.

Carotenoid pigments, in trace levels have been reported in some rice varieties. The major brown rice carotenoids are *β*-carotene and lutein (both ca. 100 ng/g) and zeaxanthin (ca. 30 ng/g) [22].

**Health Promoting Properties of Rice**

The bioactive compounds in rice have many health promoting properties. Studies have reported that rice bioactives exhibit antioxidant, anti-inflammatory, anti-tumor, and anti-diabetic properties [8, 23 - 25]. The tocotrienols and tocopherols in rice act as free radical scavengers [24]. Red yeast rice exhibits DPPH (2, 2-diphenyl-1-picrylhydrazyl) radical scavenging activity. Moreover, the germinated brown rice and red yeast rice prevents hydrogen peroxide radical induced oxidative damage to neuroblastoma cells [26, 27].

Rice bran was found to lower plasma triglyceride concentrations and increase the ratios of HDL (high density lipoprotein) cholesterol to total cholesterol and of apolipoprotein A-I to B [28] that are powerful predictors of cardiovascular disease. γ-Oryzanol has high antioxidant activity and blood cholesterol lowering properties. It also decreases cholesterol absorption [29] and platelet aggregation [29 - 31]. Makyen and co-workers [32] also reported that γ-oryzanol plays an important role in inhibiting cholesterol uptake into epithelial cells and inhibition of 3-hydroxy-3-methylglutaryl-coenzyme A (HMG-CoA) reductase, a key enzyme in the cholesterol biosynthesis pathway. Tocotrienol exhibits antiangiogenic activity [33].

The functional bioactives in stabilised rice bran were found to inhibit three key pro-inflammatory enzymes (cyclooxygenase 1, cyclooxygenase 2, and 5-lipoxygenase) that control the inflammatory cascade involved in impaired joint health, pain, and arthritis [34]. Islam and co-workers [35] studied the anti-inflammatory effect of γ-oryzanol in colitis induced rats and found that the phytosteryl ferulates showed anti-inflammatory effect which may be mediated by inhibition of NF-κB activity. Moreover, γ-oryzanol also shows immunosti-mulatory properties through cellular and humoral mediated mechanisms [36].

Cyanidin-3-glucoside extracted from black rice improves hyperlipidaemia induced by a high fat/cholesterol diet in rats in part by modulating the activities of hepatic lipogenic enzymes [37]. Red rice varieties from Sri Lanka have been reported to have antiglycation, amylase inhibition and glycation reversing properties [38].

Germinated brown rice exhibited improved antioxidant status, and improved glycaemia and kidney functioning in type-2 diabetic rats [39]. Another epidemiological study had reported that germinated brown rice exhibits reduction in the risk of some chronic degenerative diseases like cancer, cardiovascular and Alzheimer's disease [40].

Ferulic acid, tricin, *β*-sitosterol, γ-oryzanol, tocotrienol, tocopherols, and phytic acid present in rice bran have the ability to induce cell apoptosis and inhibit cell proliferation in malignant cells [41, 42]. The sterols in Thai black rice are more potent inhibitors of colon, breast and blood cell cancers than cyanidin 3-glucoside and peonidin 3-glucoside [41]. Rice varieties differ in their inhibitory effect on the colorectal cancerous cell growth [43]. Total phenolics and γ- tocotrienol were positively correlated with reduced colorectal cancerous cell growth. Colorectal cancer chemoprevention is attributed to scavenging of free radicals, blocking of inflammatory responses, modulation of gut microflora communities and regulation of carcinogen-metabolising enzymes by the bioactives in rice bran [44]. Rice bran has been suggested to be beneficial as a putative chemopreventive intervention in humans with intestinal polyps [45].

Njavara black (NB) rice bran contains three important compounds namely, tricin and two rare flavonolignans- tricin 4′-O-(*erythro-β*-guaiacylglyceryl) ether and tricin 4′-O-(*threo-β*-guaiacylglyceryl) ether. The $EC_{50}$ values of these compounds in DPPH system were 90.39, 352.04 and 208.1 µg/ml, respectively. Of the three compounds, tricin and the *threo-* form of flavonolignan showed anti-inflammatory effect of >65% after 5 h at 2 mg/kg in carrageenan-induced paw edema experiments in rats [11].

Ethyl acetate extract of brown rice showed anti-hypertensive properties by hindering the stimulatory effect of angiotensin II on rat smooth muscle cells under laboratory conditions [46]. Other *in vitro* studies had reported that pigmented rice can suppress tumor progression [41, 42]. The 70% ethanol-water extracts of bran of pigmented rice cultivars of Korea strongly inhibited phorbol-ester induced tumor promotion in marmoset lymphoblastoid cells B95-8 *in vitro* [42]. Chiang *et al.* [47] observed in their cellular study on HepG2 cells that cyanidin-3-O-glucoside and peonidin-3-O-glucoside present in the water extract of whole black rice contributed to the marked increase in superoxide dismutase (SOD) and catalase (CAT) activities. Their *in vivo* study on C57BL/6 mice revealed that feeding of the extract significantly increased plasma HDL-cholesterol and lowered thiobarbituric acid-reactive substances and enhanced SOD and CAT activities.

Rice anthocyanin extract could increase the regeneration of rhodopsin [48]. Black rice anthocyanins prevented retinal photochemical damage and inhibited retinal cell apoptosis induced by fluorescent light in Sprague Dawley rats [49]. Another study on animal model showed positive correlation between consumption of pigmented fraction of rice on the blood lipid levels and related diseases. When supplemented in diet with the pigmented rice, red and black rice showed reduction in atherosclerosis plaques formation in hypercholesterolemic rabbits [50], with 50% lower area of atherosclerotic plaque than those fed the white rice diet.

Cai *et al.* [51] observed considerable growth-inhibitory potency of tricin from rice bran in human-derived malignant MDA-MB-468 breast tumour cells, but could not ascertain the breast cancer chemoprevention ability.

### Effect of Processing on the Bioactive Compounds in Rice

Processing of rice includes milling, cooking, steaming, parboiling, fermentation, and germination. Milling and polishing treatments result in the loss of the bran layer which contains the bulk portion of the bioactive compounds in rice [52]. Cooking drastically reduces the phytochemicals and antioxidant capacities. These properties further decreases with keeping time of the cooked rices; however, retention is more in pigmented rice [53].

Extractable content of tocopherols, tocotrienols, and oryzanol increases on cooking of brown rice [54] probably due to the release from their bound forms as well as concentration of lipophilic matter subsequent to the leaching out of water soluble matter. Steaming, a common processing treatment given to rice also alters the composition and activity of bioactive compounds in rice and rice bran. Steam stabilised bran showed an increase in oryzanol, tocotrienols and tocopherols compared to the native bran, whereas, soluble polyphenols, bound polyphenols and total polyphenols and free radical scavenging activity decreased [55]. Pressure cooking of black rice resulted in ~ 80% loss of cyanidin-3-glucoside with concomitant production of protocatechuic acid, its degradation product [56]. Parboiling treatment given to paddy adversely affected the tocopherol and tocotrienol contents but had no effect on oryzanol [57]. Retention of bioactives on steaming and parboiling was more in pigmented rice [55]. Parboiling process significantly increased the concentrations of tocotrienols, tocopherols and γ-oryzanol in rice cultivars differing in bran colour but decreased the concentrations of soluble phenolic and flavonoid compounds, especially the anthocyanins and proanthocyanidins [57]. While the retention of oryzanol was 60% after parboiling, storing and cooking, only 10% of the total tocols were retained [58]. In pigmented rice, parboiling decreased the extractable anthocyanins but showed a high content of simple phenolics [57]. Parboiling reduces carotenoid contents ($\beta$ carotene, lutein, zeaxanthin) of brown rice to trace levels [22]. Parboiling of paddy using pressurised steam [59] causes decrease of bioactive compounds.

Another processing method that enhances the bioactive content in rice is fermentation. The ethanol extract of the *P. eryngii*-fermented rice showed strong reducing ability with $EC_{50}$ value less than 0.25 mg/ml [60]. Solid state fermentation with *Cordyceps sinensis* of stale rice exhibited increase in the level of ergosterol and increased superoxide dismutase activity [61]. The increase in the bioactive properties in fermented rice was attributed to the release of the bound phenolic compounds during microbial hydrolysis. In addition to that, fermentation also causes structural breakdown of the cellular matrix leading to release or formation of various phenolic compounds with radical scavenging and metal chelation properties. However, depending on the fermentation process, rice variety and the microbial culture involved, the bioactivity of the final fermented product varies [62]. Germination of rice causes an increase in bioactive compounds and antioxidant activities [63] and an increase in $\gamma$-aminobutyric acid (GABA) [64].

## CONCLUDING REMARKS

Rice, especially the bran fraction contains good amount of bioactive compounds. Compared to the non pigmented varieties, the pigmented rice has more health promoting properties. Traditional rice has many therapeutic properties in addition to being a staple food to more than half of the world population. These bioactive compounds in rice have health promoting properties such as anti-diabetic, antitumor, anti-inflammatory, antioxidant and cholesterol lowering activities. To maximise the intake of the bioactive compounds that are concentrated in the bran, consumption of rice bran and whole rice needs to be advocated. Rice bran and whole rice grain should be converted into nutritious functional foods.

## CONFLICT OF INTEREST

The authors confirm that they have no conflict of interest to declare for this publication.

## ACKNOWLEDGEMENTS

Declared none.

## REFERENCES

[1]     Goufo P, Ferreira LMM, Henrique Trindade H, Rosa EAS. Distribution of antioxidant compounds in the grain of the Mediterranean rice variety Ariete CyTA-Journal of Food 2015; 13(01): 140-50.
[http://dx.doi.org/10.1080/19476337.2014.923941]

[2]     Deepa G, Singh VK, Akhilender Naidu K. Characterization of antioxidant compounds and antioxidant activity of Indian rice varieties. J Herbs Spices Med Plants 2012; 18: 18-33.
[http://dx.doi.org/10.1080/10496475.2011.644655]

[3]     Goufo P, Trindade H. Rice antioxidants: phenolic acids, flavonoids, anthocyanins, proanthocyanidins, tocopherols, tocotrienols, γ-oryzanol, and phytic acid. Food Sci Nutr 2014; 2(2): 75-104.
[http://dx.doi.org/10.1002/fsn3.86] [PMID: 24804068]

[4]     Hirawan R, Diehl-Jones W, Beta T. Comparative evaluation of the antioxidant potential of infant cereals produced from purple wheat and red rice grains and LC-MS analysis of their anthocyanins. J Agric Food Chem 2011; 59(23): 12330-41.
[http://dx.doi.org/10.1021/jf202662a] [PMID: 22035073]

[5]     Seo WD, Kim JY, Song YC, Cho JH, Jang KC, Han SI. Comparative analysis of physicochemicals and antioxidative properties in new red rice (*Oryza sativa* L. cv. Gunganghongmi). J Crop Sci Biotechnol 2013; 16: 63-8.
[http://dx.doi.org/10.1007/s12892-012-0057-3]

[6]     Burlando B, Cornara L. Therapeutic properties of rice constituents and derivatives (*Oryza sativa* L.): A review update. Trends Food Sci Technol 2014; 40: 82-98.
[http://dx.doi.org/10.1016/j.tifs.2014.08.002]

[7]     Ahuja U, Ahuja SC, Thakrar R, Singh RK. Rice -a nutraceutical. Asian Agrihist 2008; 12: 93-108.

[8]     Rahman S, Sharma MP. Nutritional and medicinal values of some indigenous rice varieties. Indian J of Traditional Knowl 2006; 5(4): 454-8.

[9]     Deng GF, Xu XR, Zhang Y, Li D, Gan RY, Li HB. Phenolic compounds and bioactivities of pigmented rice. Crit Rev Food Sci Nutr 2013; 53(3): 296-306.
[http://dx.doi.org/10.1080/10408398.2010.529624] [PMID: 23216001]

[10]    Ma J, Li Y, Ye Q, *et al.* Constituents of red yeast rice, a traditional Chinese food and medicine. J Agric Food Chem 2000; 48(11): 5220-5.
[http://dx.doi.org/10.1021/jf000338c] [PMID: 11087463]

[11]    Mohanlal S, Parvathy R, Shalini V, Helen A, Jayalekshmy A. Isolation, characterization and quantification of tricin and flavonolignans in the medicinal rice Njavara (*Oryza sativa* L.), as compared to staple varieties. Plant Foods Hum Nutr 2011; 66(1): 91-6.
[http://dx.doi.org/10.1007/s11130-011-0217-5] [PMID: 21373805]

[12]    Hegde S, Yenagi NB, Kasturba B. Indigenous knowledge of the traditional and qualified ayurveda practitioners on the nutritional significance and use of red rice in medications. Indian J Tradit Knowl

2013; 12(4): 506-11.

[13] Rohrer CA, Siebenmorgen TJ. Nutraceutical concentrations within the bran of various rice kernel thickness fractions. Biosystems Eng 2004; 88(4): 453-60.
[http://dx.doi.org/10.1016/j.biosystemseng.2004.04.009]

[14] Huang SH, Ng LT. Quantification of tocopherols, tocotrienols, and γ-oryzanol contents and their distribution in some commercial rice varieties in Taiwan. J Agric Food Chem 2011; 59(20): 11150-9.
[http://dx.doi.org/10.1021/jf202884p] [PMID: 21942383]

[15] Adom KK, Liu RH. Antioxidant activity of grains. J Agric Food Chem 2002; 50(21): 6182-7.
[http://dx.doi.org/10.1021/jf0205099] [PMID: 12358499]

[16] Begum A, Goswami A, Chowdhury PA. Comparative study on free and bound phenolic acid content and their antioxidant activity in bran of rice (Oryza sativa L.) cultivars of eastern Himalayan range. Int J Food Sci Tech 2015; 50(12): 2529-36.

[17] Walter M, Marchesan E. Phenolic compounds and antioxidant activity of rice. Braz Arch Biol Technol 2011; 54: 371-7.
[http://dx.doi.org/10.1590/S1516-89132011000200020]

[18] Nakornriab M, Sriseadka T, Wongpornchai S. Quantification of carotenoid and flavonoid components in brans of some Thai black rice cultivars using supercritical fluid extraction and high-performance liquid chromatography-mass spectrometry. J Food Lipids 2008; 15: 488-503.
[http://dx.doi.org/10.1111/j.1745-4522.2008.00135.x]

[19] Pereira-Caro G, Watanabe S, Crozier A, Fujimura T, Yokota T, Ashihara H. Phytochemical profile of a Japanese black-purple rice. Food Chem 2013; 141(3): 2821-7.
[http://dx.doi.org/10.1016/j.foodchem.2013.05.100] [PMID: 23871029]

[20] Rogers EJ, Rice SM, Nicolosi RJ, Carpenter DR, McClelland CA, Romanczyk LJ. Identification and quantitation of γ– oryzanol components and simultaneous assessment of tocopherols in rice bran oil. J Am Oil Chem Soc 1993; 70(3): 301-7.
[http://dx.doi.org/10.1007/BF02545312]

[21] Nyström L, Mäkinen M, Lampi AM, Piironen V. Antioxidant activity of steryl ferulate extracts from rye and wheat bran. J Agric Food Chem 2005; 53(7): 2503-10.
[http://dx.doi.org/10.1021/jf048051t] [PMID: 15796586]

[22] Lamberts L, Delcour JA. Carotenoids in raw and parboiled brown and milled rice. J Agric Food Chem 2008; 56(24): 11914-9.
[http://dx.doi.org/10.1021/jf802613c] [PMID: 19012405]

[23] Verschoyle RD, Greaves P, Cai H, Edwards RE, Steward WP, Gescher AJ. Evaluation of the cancer chemopreventive efficacy of rice bran in genetic mouse models of breast, prostate and intestinal carcinogenesis. Br J Cancer 2007; 96(2): 248-54.
[http://dx.doi.org/10.1038/sj.bjc.6603539] [PMID: 17211473]

[24] Kitts DD. An evaluation of the multiple effects of the antioxidant vitamins. Trends Food Sci Technol 1997; 8(6): 198-203.
[http://dx.doi.org/10.1016/S0924-2244(97)01033-9]

[25] Parrado J, Miramontes E, Jover M, Gutierrez JF, Terán LC, Bautista J. Preparation of a rice bran

enzymatic extract with potential use as functional food. Food Chem 2006; 98: 742-8.
[http://dx.doi.org/10.1016/j.foodchem.2005.07.016]

[26]    Azmi NH, Ismail N, Imam MU, Ismail M. Ethyl acetate extract of germinated brown rice attenuates hydrogen peroxide-induced oxidative stress in human SH-SY5Y neuroblastoma cells: role of anti-apoptotic, pro-survival and antioxidant genes. BMC Complement Altern Med 2013; 13: 177.
[http://dx.doi.org/10.1186/1472-6882-13-177] [PMID: 23866310]

[27]    Kwon CS. Antioxidant properties of red yeast rice (*Monascus purpureus*) extracts. J Korean Soc Food Sci Nutr 2012; 41: 437-42.
[http://dx.doi.org/10.3746/jkfn.2012.41.4.437]

[28]    Kestin M, Moss R, Clifton PM, Nestel PJ. Comparative effects of three cereal brans on plasma lipids, blood pressure, and glucose metabolism in mildly hypercholesterolemic men. Am J Clin Nutr 1990; 52(4): 661-6.
[PMID: 2169702]

[29]    Zawistowski J, Kopec A, Kitts DD. Effects of a black rice extract (Oryza sativa L. indica) on cholesterol levels and plasma lipid parameters in Wistar Kyoto rats. J Funct Foods 2009; 1(1): 50-6.
[http://dx.doi.org/10.1016/j.jff.2008.09.008]

[30]    Patel M, Naik SN. Gamma-Oryzanol from rice bran oil e a review. J Sci Indus Res 2004; 63: 569-78.

[31]    Seetharamaiah GS, Krishnakantha TP, Chandrasekhara N. Influence of oryzanol on platelet aggregation in rats. J Nutr Sci Vitaminol (Tokyo) 1990; 36(3): 291-7.
[http://dx.doi.org/10.3177/jnsv.36.291] [PMID: 2292731]

[32]    Mäkynen K, Chitchumroonchokchai C, Adisakwattana S, Failla M, Ariyapitipun T. Effect of gamma-oryzanol on the bioaccessibility and synthesis of cholesterol. Eur Rev Med Pharmacol Sci 2012; 16(1): 49-56.
[PMID: 22338548]

[33]    Matsuzuka K, Kimura E, Nakagawa K, Murata K, Kimura T, Miyazawa T. Investigation of tocotrienol biosynthesis in rice (Oryza sativa L.). Food Chem 2013; 140(1-2): 91-8.
[http://dx.doi.org/10.1016/j.foodchem.2013.02.058] [PMID: 23578619]

[34]    Roschek B Jr, Fink RC, Li D, *et al.* Pro-inflammatory enzymes, cyclooxygenase 1, cyclooxygenase 2, and 5-lipooxygenase, inhibited by stabilized rice bran extracts. J Med Food 2009; 12(3): 615-23.
[http://dx.doi.org/10.1089/jmf.2008.0133] [PMID: 19627211]

[35]    Islam MS, Murata T, Fujisawa M, *et al.* Anti-inflammatory effects of phytosteryl ferulates in colitis induced by dextran sulphate sodium in mice. Br J Pharmacol 2008; 154(4): 812-24.
[http://dx.doi.org/10.1038/bjp.2008.137] [PMID: 18536734]

[36]    Ghatak SB, Panchal SJ. Investigation of the immunomodulatory potential of oryzanol isolated from crude rice bran oil in experimental animal models. Phytother Res 2012; 26(11): 1701-8.
[http://dx.doi.org/10.1002/ptr.4627] [PMID: 22407738]

[37]    Um MY, Ahn J, Ha TY. Hypolipidaemic effects of cyanidin 3-glucoside rich extract from black rice through regulating hepatic lipogenic enzyme activities. J Sci Food Agric 2013; 93(12): 3126-8.
[http://dx.doi.org/10.1002/jsfa.6070] [PMID: 23471845]

[38]    Premakumara GA, Abeysekera WK, Ratnasooriya WD, Chandrasekharan NV, Bentota AP.

Antioxidant, anti-amylase and anti-glycation potential of brans of some Sri Lankan traditional and improved rice (*Oryza sativa* L.) varieties. J Cereal Sci 2013; 58: 451-6.
[http://dx.doi.org/10.1016/j.jcs.2013.09.004]

[39] Imam MU, Musa SN, Azmi NH, Ismail M. Effects of white rice, brown rice and germinated brown rice on antioxidant status of type 2 diabetic rats. Int J Mol Sci 2012; 13(10): 12952-69.
[http://dx.doi.org/10.3390/ijms131012952] [PMID: 23202932]

[40] Wu F, Yang N, Touré A, Jin Z, Xu X. Germinated brown rice and its role in human health. Crit Rev Food Sci Nutr 2013; 53(5): 451-63.
[http://dx.doi.org/10.1080/10408398.2010.542259] [PMID: 23391013]

[41] Leardkamolkarn V, Thongthep W, Suttiarporn P, Kongkachuichai R, Wongpornchai S, Wanavijitr A. Chemopreventive properties of the bran extracted from newly-developed Thai rice: The riceberry. Food Chem 2011; 125: 978-85.
[http://dx.doi.org/10.1016/j.foodchem.2010.09.093]

[42] Nam SH, Choi SP, Kang MY, Koh HJ, Kozukue N, Friedman M. Bran extracts from pigmented rice seeds inhibit tumor promotion in lymphoblastoid B cells by phorbol ester. Food Chem Toxicol 2005; 43(5): 741-5.
[http://dx.doi.org/10.1016/j.fct.2005.01.014] [PMID: 15778014]

[43] Forster GM, Raina K, Kumar A, *et al.* Rice varietal differences in bioactive bran components for inhibition of colorectal cancer cell growth. Food Chem 2013; 141(2): 1545-52.
[http://dx.doi.org/10.1016/j.foodchem.2013.04.020] [PMID: 23790950]

[44] Henderson AJ, Ollila CA, Kumar A, *et al.* Chemopreventive properties of dietary rice bran: current status and future prospects. Adv Nutr 2012; 3(5): 643-53.
[http://dx.doi.org/10.3945/an.112.002303] [PMID: 22983843]

[45] Verschoyle RD, Greaves P, Cai H, *et al.* Evaluation of the cancer chemopreventive efficacy of rice bran in genetic mouse models of breast, prostate and intestinal carcinogenesis. Br J Cancer 2007; 96(2): 248-54.
[http://dx.doi.org/10.1038/sj.bjc.6603539] [PMID: 17211473]

[46] Utsunomiya H, Takaguri A, Bourne AM, *et al.* An extract from brown rice inhibits signal transduction of angiotensin II in vascular smooth muscle cells. Am J Hypertens 2011; 24(5): 530-3.
[http://dx.doi.org/10.1038/ajh.2011.10] [PMID: 21331052]

[47] Chiang AN, Wu HL, Yeh HI, Chu CS, Lin HC, Lee WC. Antioxidant effects of black rice extract through the induction of superoxide dismutase and catalase activities. Lipids 2006; 41(8): 797-803.
[http://dx.doi.org/10.1007/s11745-006-5033-6] [PMID: 17120934]

[48] Matsumoto H, Nakamura Y, Tachibanaki S, Kawamura S, Hirayama M. Stimulatory effect of cyanidin 3-glycosides on the regeneration of rhodopsin. J Agric Food Chem 2003; 51(12): 3560-3.
[http://dx.doi.org/10.1021/jf034132y] [PMID: 12769524]

[49] Jia H, Chen W, Yu X, *et al.* Black rice anthocyanidins prevent retinal photochemical damage via involvement of the AP-1/NF-κB/Caspase-1 pathway in Sprague-Dawley rats. J Vet Sci 2013; 14(3): 345-53.
[http://dx.doi.org/10.4142/jvs.2013.14.3.345] [PMID: 23820171]

[50]   Ling WH, Cheng QX, Ma J, Wang T. Red and black rice decrease atherosclerotic plaque formation and increase antioxidant status in rabbits. J Nutr 2001; 131(5): 1421-6.
[PMID: 11340093]

[51]   Cai H, Hudson EA, Mann P, *et al.* Growth-inhibitory and cell cycle-arresting properties of the rice bran constituent tricin in human-derived breast cancer cells in vitro and in nude mice in vivo. Br J Cancer 2004; 91(7): 1364-71.
[http://dx.doi.org/10.1038/sj.bjc.6602124] [PMID: 15316567]

[52]   Esa NM, Ling TB, Peng LH. By-products of rice processing: an overview of health benefits and applications. J Rice Res 2013; 1: 1-11.

[53]   Saikia S, Dutta H, Saikia D, Mahanta CL. Quality characterisation and estimation of phytochemicals content and antioxidant capacity of aromatic pigmented and non-pigmented rice varieties. Food Res Int 2012; 46: 334-40.
[http://dx.doi.org/10.1016/j.foodres.2011.12.021]

[54]   Lin T, Huang S, Ng L. Effects of cooking conditions on the concentrations of extractable tocopherols, tocotrienols and γ-oryzanol in brown rice: Longer cooking time increases the levels of extractable bioactive components. Eur J Lipid Sci Technol 2015; 117: 349-54.
[http://dx.doi.org/10.1002/ejlt.201400148]

[55]   Pradeep PM, Jayadeep A, Manisha Guha M, Vasudeva Singh V. Hydrothermal and biotechnological treatments on nutraceutical content and antioxidant activity of rice bran. J Cereal Sci 2014; 60: 187-92.
[http://dx.doi.org/10.1016/j.jcs.2014.01.025]

[56]   Hiemori M, Koh E, Mitchell AE. Influence of cooking on anthocyanins in black rice (*Oryza sativa* L. japonica var. SBR). J Agric Food Chem 2009; 57(5): 1908-14.
[http://dx.doi.org/10.1021/jf803153z] [PMID: 19256557]

[57]   Min B, McClung A, Chen MH. Effects of hydrothermal processes on antioxidants in brown, purple and red bran whole grain rice (*Oryza sativa* L.). Food Chem 2014; 159: 106-15.
[http://dx.doi.org/10.1016/j.foodchem.2014.02.164] [PMID: 24767032]

[58]   Pascual CS, Massaretto IL, Kawassaki F, Barros RM, Noldin JA, Marquez UM. Effects of parboiling, storage and cooking on the levels of tocopherols, tocotrienols and γ- oryzanol in brown rice (*Oryza sativa* L.). Food Res Int 2013; 50: 676-81.
[http://dx.doi.org/10.1016/j.foodres.2011.07.013]

[59]   Dutta H, Saikia S, Mahanta CL. Preliminary investigation on bioactive potential of milling fractions of *komal chaul* of Assam processed from pigmented kola chokua paddy. Int J Sci Technol 2014; 4(3): 1-9.

[60]   Bao L, Li Y, Wang Q, *et al.* Nutritive and bioactive components in rice fermented with the edible mushroom *Pleurotus eryngii*. Mycology 2013; 4(2): 96-102.
[http://dx.doi.org/10.1080/21501203.2013.816386]

[61]   Zhang Z, Lei Z, Lü Y, Lü Z, Chen Y. Chemical composition and bioactivity changes in stale rice after fermentation with *Cordyceps sinensis*. J Biosci Bioeng 2008; 106(2): 188-93.
[http://dx.doi.org/10.1263/jbb.106.188] [PMID: 18804063]

[62]   Hur SJ, Lee SY, Kim YC, Choi I, Kim GB. Effect of fermentation on the antioxidant activity in plant-

based foods. Food Chem 2014; 160: 346-56.
[http://dx.doi.org/10.1016/j.foodchem.2014.03.112] [PMID: 24799248]

[63]  Lin YT, Pao CC, Wu ST, Chang CY. Effect of different germination conditions on antioxidative properties and bioactive compounds of germinated brown rice. BioMed Res Int 2015; 2015: 608761.
[http://dx.doi.org/10.1155/2015/608761] [PMID: 25861637]

[64]  Komatsuzaki N, Tsukahara K, Toyoshima H, Suzuki T, Shimizu N, Kimura T. Effect of soaking and gaseous treatment on GABA content in germinated brown rice. J Food Eng 2007; 78(2): 556-60.
[http://dx.doi.org/10.1016/j.jfoodeng.2005.10.036]

# SUBJECT INDEX

## A

Anticarcinogenic iii, 8, 20, 48, 49, 65, 66, 72, 80, 81, 93, 110, 167, 168, 180, 181, 185, 190, 211

Antiinflammatory 28, 115, 122

Antioxidant iii, 3, 10, 27, 28, 30, 33, 35, 36, 38, 39, 48, 49, 56, 69, 73, 76, 77, 79, 80, 86, 88, 90, 93, 96, 99, 102, 104, 105, 114, 115, 155, 160, 170, 171, 180, 181, 183, 185, 188, 190, 191, 195, 197, 198, 201, 217, 218, 220, 221, 227-233

Ascorbic Acid 4, 37, 46, 53, 54, 60, 63, 64, 82, 91, 92, 101, 109, 138, 139, 143, 146, 150, 159

## B

Bioactive compounds i, iii, 18, 27, 28, 41, 46, 48, 59, 61, 62, 66, 67, 70, 71, 84, 110, 132, 151, 155, 170, 180, 181, 183, 197, 199, 209, 210, 227, 228, 234

Bioactive peptides 15-17

Bioactivity i, iv, 27, 32, 53, 61, 62, 79, 85, 93, 134, 164, 221, 228, 233

Bioavailability 6, 7, 18, 37, 41, 42, 44, 45, 78, 79, 93, 178

Boiling 31, 37, 38, 102, 132, 145, 146, 154

*Brassicaceae* i, 27, 28, 34, 39-43

## C

Caffeine 107, 193, 212, 213, 215-219

Cancer i, iii, 3, 4, 16, 24, 26, 27, 29, 32, 33, 35, 42, 43, 50, 63, 64, 66, 79, 90, 91, 93, 96, 97, 125, 136, 138, 139, 160, 162, 164, 167, 168, 175, 177, 178, 181, 188, 221, 230, 232, 233

*Capsicum annuum* 104, 105, 109

Carbohydrates i, 49, 112, 113, 132, 134, 155, 157, 158, 182, 183, 188

Cardioprotection 49

Cardiovascular disease 22, 26, 28, 38, 50, 63, 83, 89, 158, 175, 214, 225

Carotenoids 4, 5, 11, 12, 25, 37, 39, 43, 59, 60, 62, 63, 68, 72, 73, 78, 92, 94, 95, 101, 104, 121, 149, 224, 230

Chemical composition 5, 47, 112, 113, 121, 124, 145, 147, 148, 154, 155, 157, 172, 173, 189, 199, 202, 210, 220, 233

Chemoprevention 28, 120, 122, 128, 131, 226, 227

Chlorogenic acids 199, 200, 206, 207, 217, 218

Clinical trials 49, 50, 66, 68, 71, 98

Coffee i, 180-220

Cold storage 132, 142, 143

Cruciferous 28, 29, 33, 43, 44, 47

## D

Diterpenes 180, 181, 183, 190, 191, 195, 203, 204, 206, 210, 211, 219, 220

Drying 132, 143, 148, 152, 153

## F

Fatty acids 4, 5, 11, 12, 25, 26, 38, 64, 96, 97, 105, 113, 121, 130, 132, 135, 137, 138, 145, 149, 153, 157, 161, 175

Fibre 3, 132, 135, 136, 142, 145, 146, 148, 221, 222

Flavonoids iii, 3, 7, 12, 15, 20, 23, 29, 56, 57, 65, 78, 84, 86, 115, 118, 121, 124, 127, 144, 152, 155, 165, 167, 177, 178, 223, 229

Functional food 49, 50, 110, 149, 175, 231

www.ingramcontent.com/pod-product-compliance
Lightning Source LLC
Chambersburg PA
CBHW050825220326
41598CB00006B/314